Advancement of Deep Learning and its Applications in Object Detection and Recognition

RIVER PUBLISHERS SERIES IN COMPUTING AND INFORMATION SCIENCE AND TECHNOLOGY

The "River Publishers Series in Computing and Information Science and Technology" covers research which ushers the 21st Century into an Internet and multimedia era. Networking suggests transportation of such multimedia contents among nodes in communication and/or computer networks, to facilitate the ultimate Internet.

Theory, technologies, protocols and standards, applications/services, practice and implementation of wired/wireless networking are all within the scope of this series. Based on network and communication science, we further extend the scope for 21st Century life through the knowledge in machine learning, embedded systems, cognitive science, pattern recognition, quantum/biological/molecular computation and information processing, user behaviors and interface, and applications across healthcare and society.

Books published in the series include research monographs, edited volumes, handbooks and textbooks. The books provide professionals, researchers, educators, and advanced students in the field with an invaluable insight into the latest research and developments.

Topics included in the series are as follows:-

- Artificial intelligence
- Cognitive Science and Brian Science
- Communication/Computer Networking Technologies and Applications
- Computation and Information Processing
- Computer Architectures
- Computer networks
- Computer Science
- Embedded Systems
- Evolutionary computation
- Information Modelling
- Information Theory
- Machine Intelligence
- Neural computing and machine learning
- Parallel and Distributed Systems
- Programming Languages
- Reconfigurable Computing
- Research Informatics
- Soft computing techniques
- Software Development
- Software Engineering
- Software Maintenance

For a list of other books in this series, visit www.riverpublishers.com

Advancement of Deep Learning and its Applications in Object Detection and Recognition

Editors

Roohie Naaz Mir
National Institute of Technology, India

Vipul Kumar Sharma
Jaypee University of Information Technology, India

Ranjeet Kumar Rout
National Institute of Technology, India

Saiyed Umer
Aliah University, India

NEW YORK AND LONDON

Published 2023 by River Publishers
River Publishers
Alsbjergvej 10, 9260 Gistrup, Denmark
www.riverpublishers.com

Distributed exclusively by Routledge
605 Third Avenue, New York, NY 10017, USA
4 Park Square, Milton Park, Abingdon, Oxon OX14 4RN

Advancement of Deep Learning and its Applications in Object Detection and Recognition / by Roohie Naaz Mir, Vipul Kumar Sharma, Ranjeet Kumar Rout, Saiyed Umer.

Routledge is an imprint of the Taylor & Francis Group, an informa business

ISBN 978-87-7022-702-5 (print)
ISBN 978-10-0088-041-0 (online)
ISBN 978-1-003-39365-8 (ebook master)

Contents

Preface

Object detection, as one of the most fundamental and challenging problems in computer vision, has received significant attention in recent years. Its development in the past two decades can be regarded as the epitome of computer vision history. If we think of today's object detection as a technical aesthetics under the power of deep learning, then turning back the clock 20 years, we would witness the wisdom of the cold weapon era. Object detection is an important computer vision task that deals with detecting instances of visual objects of a particular class in digital images.

This book focuses on describing and analyzing deep learning-based object detection tasks. The existing books always cover a series of domains of general object detection. Due to the rapid development of computer vision research, they may not contain state-of-the-art methods that provide novel solutions and new directions for these tasks. The book lists every novel solution proposed recently but neglects to discuss the basics so the readers can see the field's cutting edge more quickly. Moreover, different from the previous object detection books, this project systematically and comprehensively reviews deep learning-based object detection methods and, most notably, the up-to-date detection solutions and a set of significant research trends as well. The book is featured by in-depth analysis and discussion in various aspects, many of which is the first time in this field to the best of our knowledge.

The book mainly focuses on stepwise discussion, exhaustive literature review, detailed analysis and discussion, rigorous experimentation results. Moreover, an application-oriented approach is demonstrated and encouraged.

List of Figures

List of Tables

List of Contributors

Anand, Mansimar *Department of Computer Science & Engineering, Thapar Institute of Engineering and Technology, India, anandmansimar@gmail.com*

Asari, Vijayan *Electrical and Computer Engineering, University of Dayton, 300 College Park, USA, vasari1@udayton.edu*

Aviansh, D. Omkar, *E-mail: domka1@unh.newhaven.edu*

Busa, Srikanth, *Department of Computer Science & Engineering, Kallam Haranadhareddy Institute of Technology, India; E-mail: srikanth.busa@gmail.com*

Chakraborty, Chandan, *Department of Computer Science & Engineering, NITTTR, India; E-mail: chakraborty.nitttrk@gmail.com*

Dhara, Bibhas Chandra, *E-mail: bcdhara@gmail.com*

Dhibar, Somenath, *E-mail: somenath.ju@gmail.com*

Garg, Gaurish, *Department of Computer Science & Engineering, Thapar Institute of Engineering and Technology, India*

Ghosh, Mridul, *Department of Computer Science, Shyampur Siddheswari Mahavidyalaya, India; E-mail: mridulxyz@gmail.com*

Grandhe, Padmaja, *Department of Computer Science & Engineering, Potti Sriramulu Chalavadhi Mallikarjunarao College of Engineering & Technology, India; E-mail: padmajagrandhe@gmail.com*

Jana, Prithwish, *Department of Computer Science & Engg., Indian Institute of Technology Kharagpur, India; E-mail: pjana@ieee.org*

Kapoor, Navpreet Singh, *Department of Computer Science & Engineering, Thapar Institute of Engineering and Technology, India; E-mail: navpreetsingh181124@gmail.com*

Lasker, Asifuzzaman, *Department of Computer Science & Engineering, Aliah University, India; asifuzzaman.lasker@gmail.com*

Malhi, Avleen, *Data science and AI, Bournemouth university UK; E-mail: amalhi@bournemouth.ac.uk*

Mir, Roohie Naaz, *Department of Computer Science & Engineering, National Institute of Technology Srinagar, India; E-mail: naaz310@nitsri.net*

Mohanta, Partha Pratim, *Electronics and Communication Sciences Unit, Indian Statistical Institute, India; E-mail: ppmohanta@isical.ac.in*

Obaidullah, Sk Md, *Department of Computer Science & Engineering, Aliah University,India; E-mail: sk.obaidullah@gmail.com*

Pero, Chiara, *E-mail: cpero@unisa.it*

Priyanshu, *Department of Computer Science & Engineering, Thapar Institute of Engineering and Technology, Patiala, India; E-mail: priyanshuagg08@gmail.com*

Radhika, K. S. R, *E-mail: ksrradhika@tkrcet.com*

Roy, Kaushik, *Department of Computer Science, West Bengal State University, India; E-mail: kaushik.mrg@gmail.com*

Sharma, Sariva, *Dr B R Ambedkar National Institute of Technology Jalandhar, India; E-mail: sariva03@gmail.com*

Sharma, Vipul, *Department of Computer Science & Engineering, National Institute of Technology Srinagar, India; E-mail: vipul_1phd17@nitsri.net*

Shivani, Shivendra, *Department of Computer Science & Engineering, Thapar Institute of Engineering and Technology, India; E-mail: shivendra.shivani@thapar.edu*

Singh, Raman, *School of Computing, Engineering and Physical Sciences, University of the West of Scotland, United Kingdom; E-mail: raman.singh@uws.ac.uk*

Somala, Jayaprada, *Department of Computer Science & Engineering, Lakireddy Bali Reddy College of Engineering, India; E-mail: jayasomala@gmail.com*

Sri, T. Santhi, *Department of Computer Science & Engineering, Koneru Lakshmaiah Education Foundation, India; E-mail: santhisri@kluniversity.in*

Swathi, B, *E-mail: buragaddaswathi@gmail.com*

Teli, Mohammad Nayeem, *Department of Computer Science, University of Maryland, College Park, United States; E-mail: nayeem@cs.umd.edu*

Tiwari, Shailendra, *Department of Computer Science & Engineering, Thapar Institute of Engineering and Technology, India;*
E-mail: shailendra@thapar.edu

Ullah, Nasib, *École de technologie supérieure, Montreal, Quebec, Canada;*
E-mail: nasib.ullah.1@ens.etsmtl.ca

List of Abbreviations

ASNet	Attention scaling network
ASSD	Attentive single shot detection
AUC	Area under curve
BLEU-4	A method for automatic evaluation of machine translation
BoW	Bag-of-words
CAM	Context attention module
CIDEr	Consensus-based image description evaluation
CLAHE	Contrast limited adaptive histogram equalization
CMYK	Cyan, magenta, yellow, and key (black)
CNN	Convolutional neural network
COCO	Common objects in context
C-ORG	COORG relational graph
CoxPH	Cox proportional hazards
CTMF	Color texture Markov feature
CXR	Chest radiography
DANet	Density attention network
DBAS	Dynamic biometric authentication system
DBTR-BR	Discrete beamlet transforms regularization for effective blind restoration
DCGAN	Deep convolutional generative adversarial network
DCV	Discriminant common vector
DD	Daisy descriptor
DETR	Detection transformer
DL	Deep learning
DLDA	Direct linear discriminant analysis
DoG	Difference-of-Gaussian
DSIFT	Dense scale invariant feature transform
DTN	Deep tree network
DWT	Discrete wavelet transforms

EAR	Eye aspect ratio
EBGM	Elastic bunch graph matching
EMC	Ensemble monte carlo
EoT	Edginess of texture
FARET	Facial recognition technology (FERET) database
FCN	Fully convolutional network
FER	Facial expression recognition
FERS	Facial expression recognition system
FLDA	Fractional linear discriminant analysis
FPN	Feature pyramid network
FRGC	Face recognition grand challenge
FSSD	Feature fusion single shot detection
GAN	Generative adversarial network
GBM	Glioblastoma
GCN	Graph convolutional network
GLAC	Gradient local auto correlation
GLCM	Gray level co-occurrence matrix features
GME	Gabor motion energy
GPU	Graphics processing unit
HAAR	Haar wavelet
HLGAN	Hybrid leaf generative adversarial network
HOG	Histogram of oriented gradient
KDEF	Karolinska directed emotional faces
KLT	Kanade-lucas-tomasi
LBG	Linde-buzo-gray
LBP	Local binary pattern
LDA	Linear discriminant analysis
LGBP-TOP	Local gabor binary patterns from three orthogonal planes
LPP	Linear programming problem
MAE	Mean absolute error
mAP	Mean average precision
METEOR	An automatic metric for MT evaluation with improved correlation with human judgments
MFN	Multi-level feature fusion network
ML	Machine learning
MRI	Magnetic resonance imaging
MSE	Mean square error
MSVD	Microsoft video description

NUAA	Face anti-spoofing database
OA-BTG	Object-aware aggregation with bidirectional temporal graph for video captioning
OPG	Orthopantomograms
PBA	Population-based augmentation
PCA	Principal component analysis
PDE	Partial differential equation
PERCLOS	Percentage of eye closure
PGGAN	Patch global generative adversarial network
PRM	Patch rescaling module
PSNR	Peak signal to noise ratio
RaFD	Radboud faces database
RANet	Relational attention network
RBF	Redial basis function
RCNN	Region-based convolutional neural network
ReLU	Rectified linear unit
RFE	Recursive feature elimination
RGB-D	Red green blue color model
RMSE	Root mean squared error
ROC	Receiver operating characteristic
ROI	Region of interest
ROUGE-L	Recall-oriented understudy for gisting evaluation
RPN	Region proposal network
RSF	Random survival forest
RTM	Regression-token module
RT-PCR	Polymerase chain reaction
SaCNN	Scale-adaptive convolutional neural network
SANet	Scale aggregation network
SARS-CoV-2	Severe acute respiratory coronavirus 2
SGD	Stochastic gradient descent
SIFT	Scale-invariant feature transformation
SMOTE	Synthetic minority oversampling technique
SNR	Signal to noise ratio
SPPNet	Spatial pyramid pooling in CNN
SSD	Single shot detection
SSIM	Structural similarity index
STASN	Spatio-temporal anti-spoofing network
STG-KD	Spatio-temporal graph for video captioning with knowledge distillation

SURF	Speeded-up robust feature
SVM	Support vector machine
TAM	Token-attention module
THGL	Tal Hassner and Gil Levi
UAV	Unmanned aerial vehicle
VGG	Visual geometry group
VOC	Visual object class
WHO	World health organization
XGBT	eXtreme gradient boosting
YOLO	You only look once
ZSFA	Zero-shot facial anti spoofing

1

Recent Advances in Video Captioning with Object Detection

Nasib Ullah[1] and Partha Pratim Mohanta[2]

[1]École de technologie supérieure, Montreal, Quebec, Canada
[2]Electronics and Communication Sciences Unit, Indian Statistical Institute, India
E-mail: nasib.ullah.1@ens.etsmtl.ca; ppmohanta@isical.ac.in

Abstract

Object detection, a primary area of computer vision, has tremendously boosted other computer vision tasks ranging from fine-grained classification to captioning. Post Deep learning object detection methodology can be broadly segregated into two types: (i) Two-stage region proposal-based methods and (ii) Single-stage regression-based methods. In this chapter, we first overview both types of object detection methodology. However, our primary focus lies in the second part, which describes the advancements in the video captioning task due to improved object detectors.

Keywords: Object detection, object interaction features, video captioning.

1.1 Introduction

Object detection is one of the predominant areas of computer vision, where the task is to identify and locate predefined objects in images or videos. Object detection is widely used in many areas, such as autonomous driving, face detection, activity recognition, object tracking, image annotation, and many more. Methods from the pre-deep learning era use different hand-crafted features such as HOG [1], SIFT [2], LBP [3] with linear and nonlinear classifiers (SVM [4], random forest[5]). The main breakthrough in object detection happened with the help of pre-trained convolutional neural networks [6–8] and the Imagenet dataset [9]. The essential milestones in post-deep learning-based object detection are shown in Figure 1.1. RCNN [10] uses the VGG [6] network to extract features of the regions generated by selective search [11] and feed those features to an SVM classifier [4]. The extraction of features for all (around 2000) regions requires running the VGG [6] network many times. This is a computationally expensive process which makes the detection slow. SPPNet [12] introduces a spatial pyramid pooling layer to tackle the feature extraction of multiple regions quickly. Fast RCNN [13] replaces the traditional classifier with a dense layer using the region of interest pooling. Although fast RCNN [13] is better than RCNN [10], both use the same selective search [11] algorithm, making the detection pipeline slow. Faster RCNN [14] replaces the selective search [11] with region proposal network (RPN) and introduces anchor boxes. Although these methods have outperformed their predecessors in terms of performance, they are slow and not applicable to real-time applications. YOLO [15] and SSD [16] are designed for real-time application use regression. RetinaNet [17] improves performance upon YOLO [15] and SSD [16] using the focal loss function. To alleviate the quantization error, Mask RCNN [18] replaces the region of interest pooling with the region of interest align layer. Despite the improvements, the formal methods cannot detect small objects. To handle that, feature pyramid network (FPN) [19] uses feature maps at different scales and the focal loss function. Finally, DETR [20] uses the successful transformer models and set-based global loss function to detect objects in an end-to-end approach.

In the following sections, we first describe some of the state-of-the-art object detection methods using deep learning and then how these methods can be leveraged in video captioning tasks.

1.2 Post-deep Learning Object Detection

Object detection focuses on predicting predefined object classes and detecting their location in the image using bounding boxes. The most straightforward approach is to label every pixel or label for every window in a sliding window mechanism. It might be a straightforward approach but computationally costly. The breakthrough happened with the inception of region proposal-based two-stage methods. Over time, two-stage methods have reached superior performance, but one shortcoming is their slow inference time.

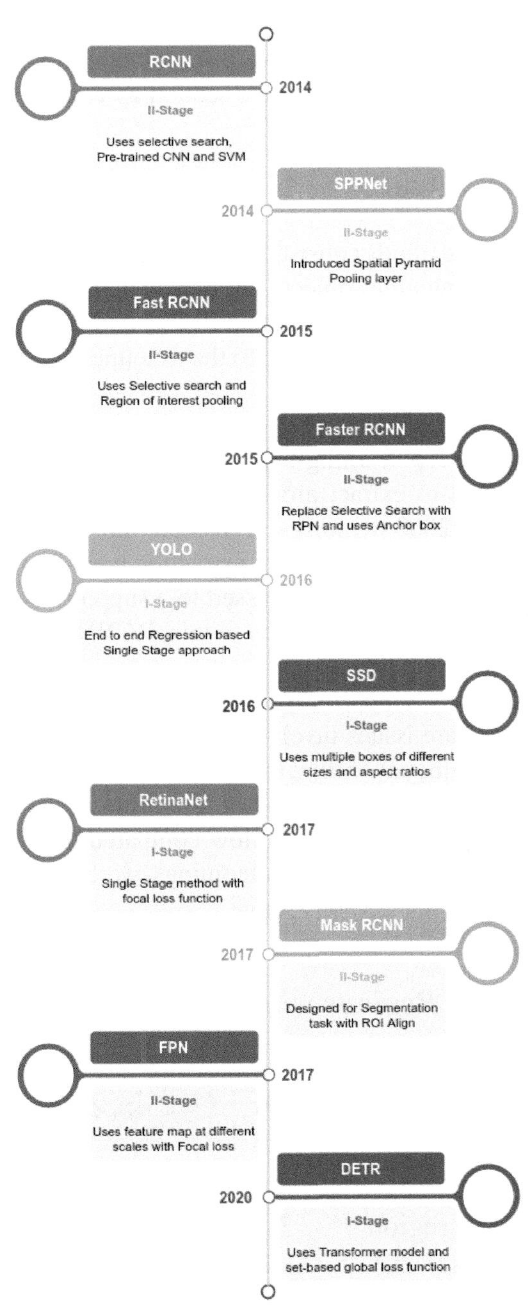

Figure 1.1 Essential milestones in object detection in post-deep learning era.

Therefore, single-stage methods are designed to tackle real-time applications and speed up the inference time. Below we discuss both group of methods and their use cases. Finally, we conclude this section by comparing different object detection methods.

1.2.1 Region proposal-based methods (Two stages)

Regional proposal-based methods consist of two stages. In the first stage, the most probable regions containing objects are detected. Then, the proposed regions are classified for the specific class in the second stage. We discuss below the most popular methods according to the timeline.

1.2.1.1 RCNN

RCNN [10] is the first deep learning-based object detection method that uses selective search [11] to extract around 2000 regions of interest from the image. Then, these 2000 regions are passed through a pre-trained convolutional neural network for the feature extraction (step-3 in Figure 1.2). Finally, the extracted region features are passed to a support vector machine [4] classifier to predict the class. In addition to this, RCNN also detects four offset values to refine the bounding boxes.

The results significantly outperform previous non-deep learning-based methods. However, there are issues involved in the methodology of RCNN: (i) it takes a considerable amount of time to extract features and classify 2000 proposed regions and (ii) the region proposal algorithm selective search is fixed, and there is no learning. Also, it is slow compared to the other deep learning counterparts. Some of these shortcomings were addressed in the follow-up fast RCNN model described below.

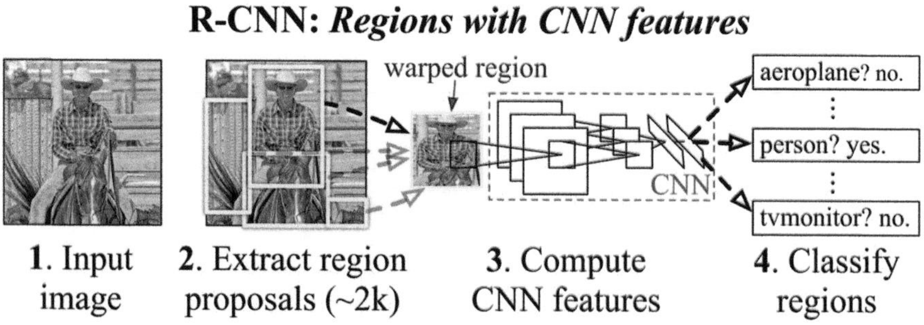

Figure 1.2 Steps for object detection in RCNN [10].

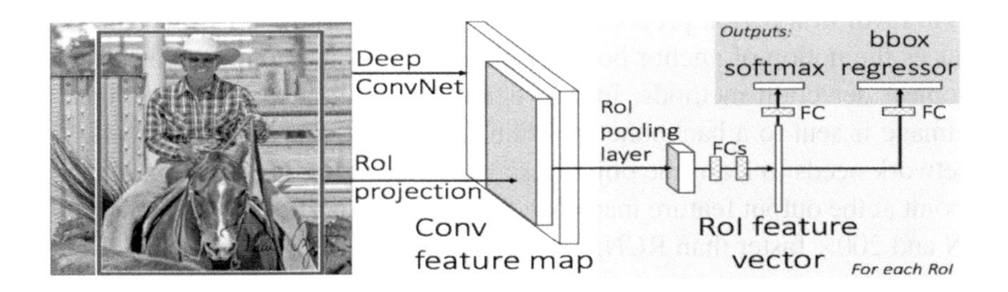

Figure 1.3 Steps for object detection in fast RCNN [13].

1.2.1.2 Fast RCNN

The same creator of RCNN has fixed some of the flaws in the second stage of RCNN and termed the new approach fast RCNN [13]. In this method, instead of feeding every region proposal separately to a convolutional neural network, the entire image is passed to generate a whole convolutional feature map. As a result, the proposed region in the convolutional feature maps will be of different shapes, but the classifier expects a fixed-size vector. To address that, region of interest (ROI) pooling is introduced. Region of interest (ROI) uses max pooling strategy to turn ununiform features.

Although fast RCNN is quicker than RCNN, it retains the same selective search technique for region proposals. As a result, the model is not trainable end-to-end. These difficulties are addressed in the following faster RCNN model.

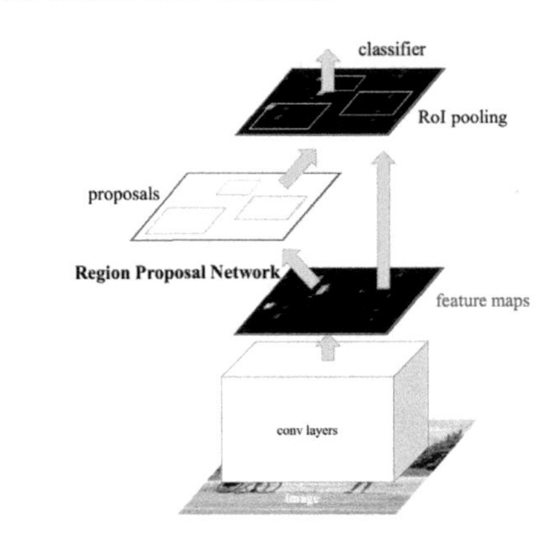

Figure 1.4 Overview of steps for faster RCNN. [14].

1.2.1.3 Faster RCNN

Both RCNN and the fast RCNN algorithms propose regions via selective search.

However, selective search is inefficient and degrades network performance. The faster RCNN [14] algorithm does away with the selective search in favor of a region proposal network (RPN). Additionally, this study introduces the notion of anchor boxes, which becomes the gold standard for most object detection methods. In the region proposal network (RPN), the input image is sent to a backbone pre-trained convolutional neural network. The network needs to learn the objectness score and bounding box offset for each point at the output feature map. The faster RCNN is 10× faster than fast RCNN and 200× faster than RCNN.

1.2.1.4 Mask RCNN

Mask RCNN [18] is an instance image segmentation method based on faster RCNN and leveraged by other vision tasks (for enhanced features) as frequently as object detection methods. It is state of the art in terms of instance segmentation. Image segmentation in computer vision is the task of partitioning a digital image into multiple segments (sets of pixels, also known as image objects). This segmentation technique is used to determine the location of objects and boundaries. Mask RCNN is built on top of faster RCNN. As discussed earlier, faster RCNN has two outputs corresponding to each candidate object (proposed by RPN), bounding box offset and class label. In contrast, mask RCNN contains an additional branch that outputs object mask.

As shown in Figure 1.5, the input image is first run through the backbone CNN, and the extracted feature maps are sent to the region proposal network (RPN). RPN along with non-max suppression (to reduce the noisy regions) provide the region of interest. In the case of faster RCNN, the proposed regions are passed through a region of interest pooling layer whose function is to transform the variable size bounding boxes into a fixed-size vector. Unfortunately, the ROI pooling layer mechanism suffers from quantization error, and to alleviate that, mask RCNN uses ROI align instead. Finally, the output feature vectors are sent to the corresponding heads to predict class probability, bounding-box offset, and mask.

1.2.2 Single-stage regression-based methods

Two-stage region proposal-based methods do not look at the whole image. Instead, it looks for an object in the proposed regions. On the other hand, single-stage methods use a single convolutional network that predicts the bounding boxes and their class probability. As discussed below, the most popular single-stage detector methods are YOLO [15] and SSD [16].

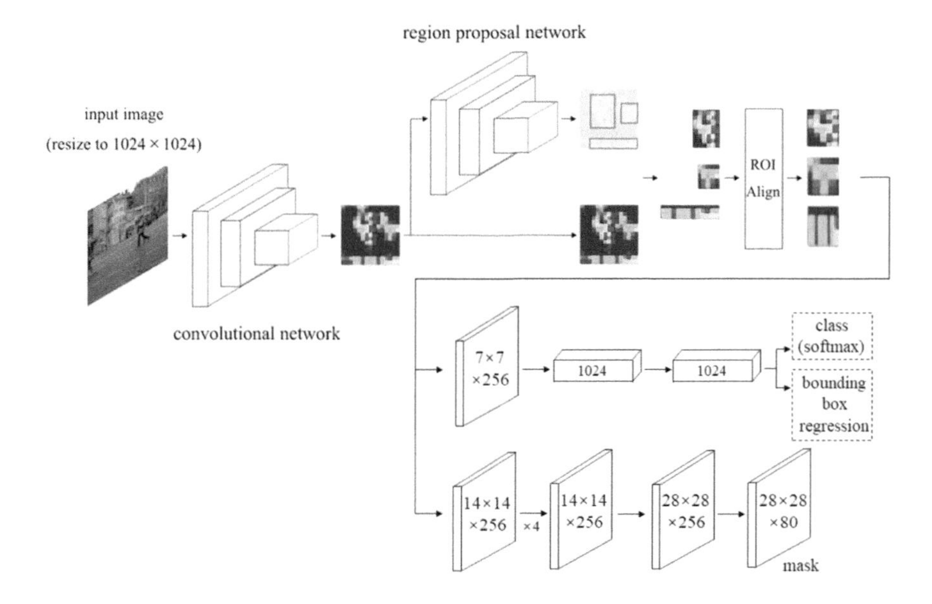

Figure 1.5 Overview of the mask R-CNN pipeline for segmentation [18].

1.2.2.1 You only look once (YOLO)

In YOLO [15], the input image is divided into S×S grids, and each grid consists of m bounding boxes. The network predicts the class probability and offsets values for each grid. So the network, which is a fully convolutional network, predicts a multidimensional tensor that encodes the global image (or for S×S grids). The bounding boxes with a class probability of more than a threshold value are selected and used to locate the object.

YOLO is 1000× faster than RCNN and 100× faster than fast RCNN, making it suitable for real-time detection applications such as self-driving automobiles. However, YOLO is still inferior in terms of performance than two-stage region proposal-based approaches.

1.2.2.2 Single-shot detection (SSD)

Like YOLO, single-shot detection (SSD) [16] also considers global images and takes only one shot to detect multiple objects. However, SSD is more accurate than YOLO, which is achieved by multiple boxes or filters of different sizes and aspect ratios. These filters are applied to multiple feature maps at the later stage of the network, enable to detect at multiple scales.

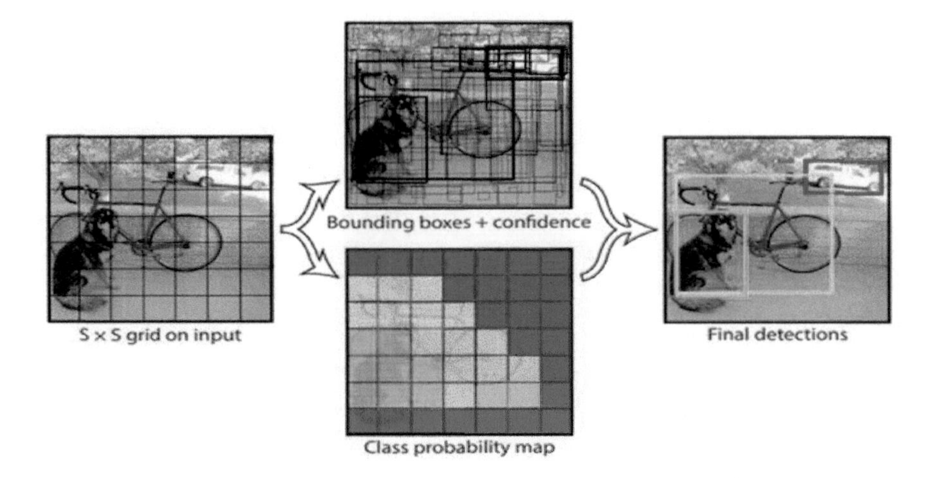

Figure 1.6 Detection using YOLO [15].

As shown in Figure 1.7, to enable predictions at many scales, the SSD output module gradually downsamples the convolutional feature maps, generating bounding box predictions on an intermittent basis. One significant difference between SSD and YOLO is that SSD does not predict the value of p_{obj}, whereas the YOLO model predicts the object probability first and then predict the class probability given there was an object present. The SSD model aims to predict the probability of a class being present in a given bounding box directly.

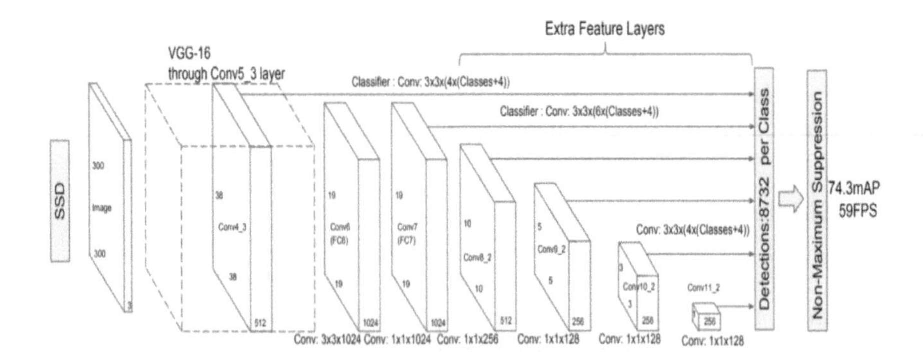

Figure 1.7 Single shot detection [16].

1.2.3 Comparison of different object detectors

We have shown the performance and inference speed comparison of the described object detection methods. The performance comparison is made using the mean average precision (mAP) metric on PASCAL VOC [21] dataset. The dataset contains around 10k images (5k trainval and 5k test) over 20 object categories. The average precision (AP) in mAP is the precision averaged across all recall values between 0 and 1. The comparisons are shown in Table 1.1.

1.3 Video Captioning with Object Detection

The objective of video captioning is to describe the content of a video using natural language. In the subsequent sections, we introduce the standard video captioning framework and then methods that leverage object detectors to improve the performance. Finally, we conclude with the performance comparison between object detector-based and non-object detector-based methods.

1.3.1 Encoder–decoder-based framework

The video captioning problem requires generating a sentence $S = \{w_1, w_2, \ldots, w_T\}$ that represents the content of a given video $V = \{v_1, v_2, \ldots, v_N\}$, where vi is a video frame and wi is a caption word. The standard encoder–decoder paradigm generates the distribution of conditional words based on preceding tokens and visual features.

$$P(S|V;\theta) = \sum_{t=1}^{T} \log P(w_t|V, w_{t-1};\theta), \tag{1.1}$$

Table 1.1 Performance and runtime comparison of different object detectors. The mAP is measured in the PASCAL VOC dataset and FPS stands for frames per second.

Model	mAP (%)	FPS	Real-time speed
RCNN [22]	66	0.1	No
Fast RCNN [23]	70	0.5	No
Faster RCNN [24]	73.2	6	No
Mask RCNN [25]	78.2	5	No
YOLO [26]	63.4	45	Yes
SSD [26]	72.4	59	Yes

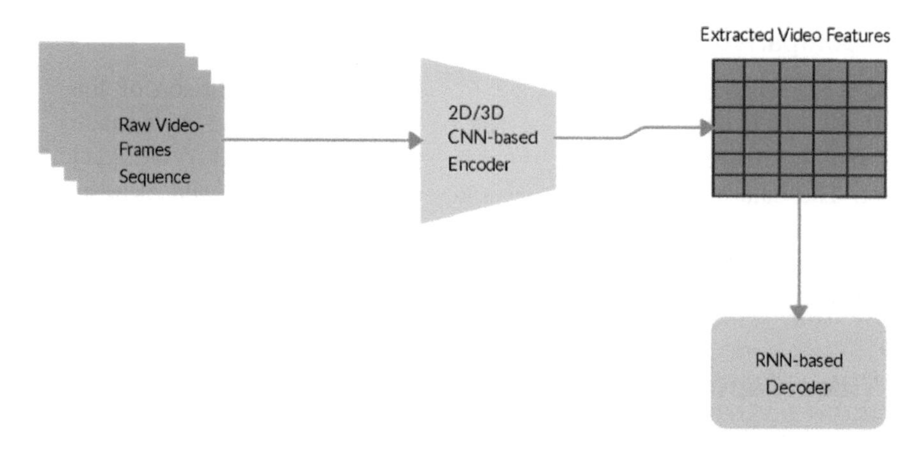

Figure 1.8 Encoder–decoder framework for video captioning.

where θ is the learnable parameters and w_t denotes the tth word in the sentence S of length T.

Encoder. To generate an appropriate caption, semantically rich visual cues must be retrieved. An encoder is in charge of extracting video features. The downstream decoder receives the extracted visual features. A common encoding technique is to extract video frame features using pre-trained 2D ConvNets (Inception-v4 [7], ResNet [8], and so on) and merge them to get a fixed-length video feature.

Decoder. The decoder creates the caption word by word using the encoder's visual features. Due to its greater temporal modeling capacity, a recurrent neural network is used as the decoder's backbone.

The encoder and decoder models are trained together by maximizing the log-likelihood of the sentence given the corresponding video feature.

$$\theta^* = \underset{\theta}{\text{argmax}} \sum_{V,S} P(S|V;\theta). \tag{1.2}$$

1.3.2 Methods with object detection

The conventional encoder–decoder approaches make use of two types of visual features: (i) global frame features taken from pre-trained convolutional neural networks and (ii) motion features (either optical flow or 3D convolutional network features). These two features are combined and passed

Figure 1.9 Illustration of the significant objects denoted by colored rectangular boxes and a sample temporal trajectory shown by a dashed curve, which are required for discriminative video understanding in order to create appropriate captioning descriptions [27].

to the decoder to generate captions. With advances in object detection and its improved performance in the open data, video captioning can now use object features. The most straightforward object-related visual features can be calculated by averaging all objects feature vectors (corresponding to each bounding box). The following sections present many state-of-the-art approaches for video captioning that use object detection.

1.3.2.1 Object-aware aggregation with bidirectional temporal graph for video captioning (OA-BTG)

According to Zhang *et al.* [27], it is crucial for video captioning to capture significant entities with accurate temporal trajectory details and represent them using discriminative spatio-temporal features. For example, the reference captioned statement "A Chinese guy throws a basketball into the basket," as shown in Figure 1.9, includes three key entities in the video: the basketball, the man and the basket, all of which need object-aware video comprehension. Furthermore, the caption description also defines the activity that the youngster is executing, "shoots a basketball," that necessitates an understanding of the boy's and the basketball's exact temporal dynamics.

Zhang *et al.* [27] used a bidirectional temporal graph to describe extensive temporal trajectories for the significant entities. The bidirectional temporal graph is built based on the similarities between object areas in various frames. It includes a forward and backward graph to get entity trajectories in both directions in the temporal order. In the input video V for each frame v_t,

N objects are extracted $R^{(t)} = \{r_i^{(t)}\}$, where $i = 1, 2, ..., N$, $t = 1, 2, ..., T$, and T is the number of sampled frames of V. The object areas in v_1 frame are used as anchors to measure the commonalities with object areas in other frames for the generation of the forward graph, with the similarity score indicating whether the two object regions correspond to the identical object instance. The backward graph is built similarly with the object areas in frame v_T serving as anchors for computing similarity scores. The bidirectional temporal graph is then created by combining the forward and backward graphs.

To link multiple instances of the same object in various frames, object areas in other frames are compared to the anchor object areas and then aligned using the nearest neighbor (NN) method to the anchor object areas. Specifically, for the object area $r_i^{(1)}$ in anchor frame v_1 and N object areas $R^{(t)} = \{r_j^{(t)}\}$ in frame v_t , $t = 2, ..., T$, the object region $\mathrm{argmax}_{r_j^{(t)}}(S(i,j))$ is aligned to the object region $r_i^{(1)}$, which means they are considered to belong to the same object instance. $S(i,j)$ is the similarity distance between ith and jth object. Also, using the same NN strategy, object areas in other frames are aligned to the anchor object areas in v_T frame (the final frame). There are $2N$ groups of object areas in the alignment process. The forward graph's N groups are then arranged in temporal order to acquire forward temporal trajectories of identified object instances. At the same time, the backward graph's N groups are arranged in temporal order to yield backward temporal trajectories. Finally, the object-aware temporal trajectory is combined with other features and sent to the decoder for caption generation.

1.3.2.2 Object relational graph with teacher recommended learning for video captioning (ORG-TRL)

The prior technique explored the temporal trajectory property of salient objects, while Zhang *et al.* [28] employed a graph neural network to represent object interaction. However, the object-relational features are more meaningful and help to achieve superior performance than non-object detection-based video captioning methods.

The pre-trained object detector is used to collect different class-agnostic entity suggestions in each keyframe and retrive their characteristics to get comprehensive entity representations $R^i = \{r_k^i\}, i = 1, ..., L, k = 1, ..., N$, where r_k^i denotes the kth entity feature in the ith keyframe, L denotes the

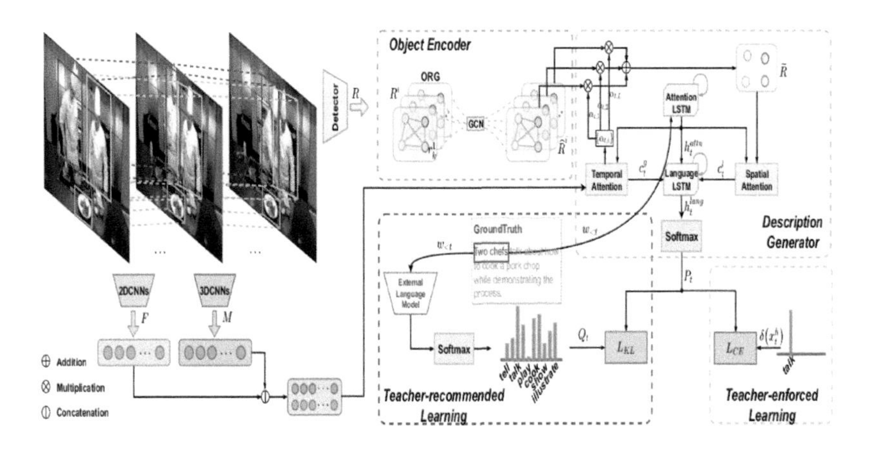

Figure 1.10 The following is a high-level summary of the ORG-TRL [28]. It is composed mostly of an object encoder (top-left box) and a hierarchical decoder with temporal/spatial attention (top-right box) [28].

keyframe count, and N specifies the total number of entities in each frame. These initial object characteristics are self-contained and do not interact in time or space. A relational graph is constructed for a group of objects to learn the relation message from surrounding objects and then update the object characteristics. Each object is treated as a node given a set of K items. Let $R \in \mathbb{R}^{K \times d}$ indicate K object nodes with d dimensional features, and $A \in \mathbb{R}^{K \times K}$ signify the K node relation coefficient matrix. A is defined as

$$A = \Phi(R).\Psi(R)^T \tag{1.3}$$

$$\Phi(R) = R.U + b_1, \Psi(R) = R.V + b_2, \tag{1.4}$$

where $U, V \in \mathbb{R}^{d \times d}$ and $b_1 \in R^d, b_2 \in R^d$ are learnable parameters. A is normalized so that the sum of edges connecting to a single node is equal to 1.

$$\hat{A} = softmax(A), \tag{1.5}$$

where $\hat{A}$ is the amount of information the center item receives from its surroundings. GCNs are used to do relational reasoning, and then the original features of R are modified to $\hat{R}$:

$$\hat{R} = \hat{A}.R.W, \tag{1.6}$$

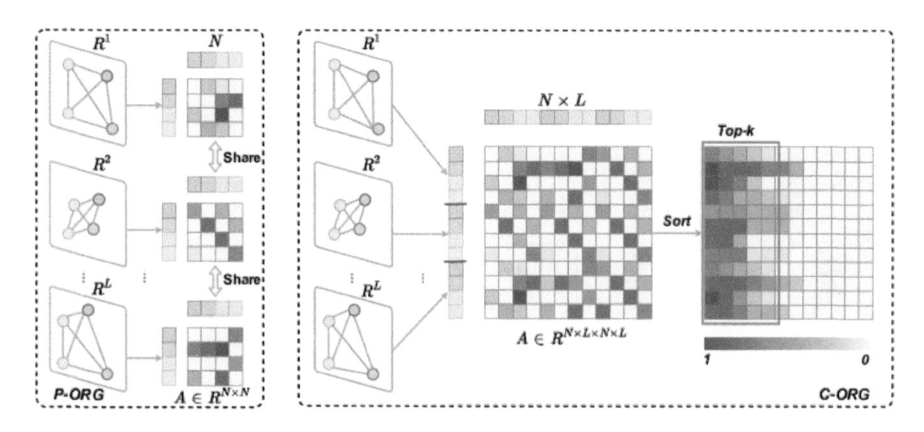

Figure 1.11 The designed C-ORG and P-ORG diagrams. Each colored square denotes the object's vector. A is the matrix of relational coefficients [28].

where $\hat{R} \in \mathbb{R}^{K \times d}$ is enriched entity features with interaction message between entities, and $W \in \mathbb{R}^{d \times d}$ is trainable parameters.

As shown in Figure 1.11, there are two kinds of relational graphs, the C-ORG and the P-ORG. The P-ORG forms relationships between N items in the same frame, resulting in an $A \in \mathbb{R}^{N \times N}$ relational graph. It is worth noting that all L frames share the learnable parameters of the relational graph. Even though object proposals arriving in various frames may be from the same entity, they are treated as separate nodes due to their various statuses. The C-ORG, on the other hand, creates a full graph $A \in \mathbb{R}^{(N \times L) \times (N \times L)}$ that links each item to all of the other $N \times L$ objects in the video. To avoid noise, only top-k matching nodes are connected. Finally, the object-relational features are combined with appearance and motion features and sent to the decoder.

1.3.2.3 Spatio-temporal graph for video captioning with knowledge distillation (STG-KD)

This method proposed by Pan *et al.* [29] generates object interaction features over time. It also suggests an object-aware knowledge distillation process, in which local object information is employed to regularise global scene features, in order to prevent unstable performance caused by a variable number of objects. Due to the variable quantity of objects in videos, the learnt object representation is often noisy. As a result, performance is unsatisfactory. To address this issue, a two-branch network topology is proposed in which an object branch captures privileged information about entity interaction and

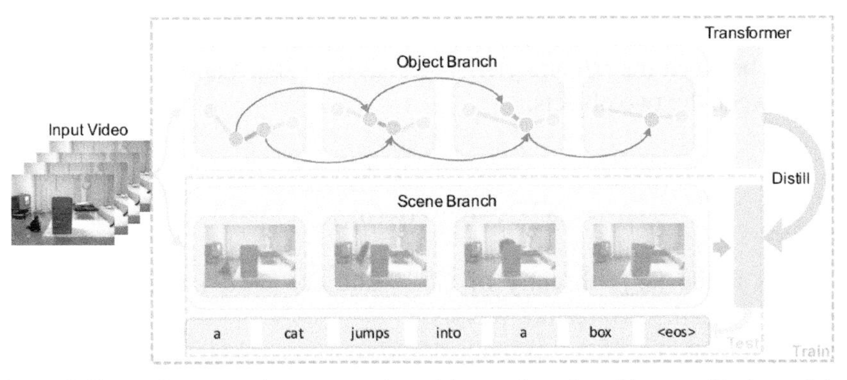

Figure 1.12 A diagrammatic representation of the two-branch architecture. During training, the object branch utilizes the proposed spatio-temporal graph model to collect space-time object interaction information, while the scene branch supplies the global context that the object branch lacks. By aligning language logits from the two branches, the object-level information is subsequently distilled into the scene feature representation [29].

subsequently injects it into a scene branch through knowledge distillation between their language logits.

As shown in Figure 1.12, the input video is initially represented as a spatio-temporal network, with nodes representing entities and edges indicating their associations. To ensure that the links are interpretable and meaningful, the adjacency matrices are constructed in such a way that they explicitly include previous information about the spatial layout and temporal transformation. Following that, the graph representation is updated by graph convolution. This modified representation is then injected into another scene branch using the suggested object-aware knowledge distillation process.

Objects behave quite differently in various regions of space and time. Distinct objects spatially interact with one another, also, the same objects undergo temporal transformations (shape, position, attitude, etc.). To account for these two distinct forms of interactions, the graph is divided into two components: the spatial and temporal graphs. The spatial graph is constructed for each frame with the adjency matrix G_t^{space} for time step t. For the temporal graph, only edges between adjacent frame pair is calculated rather than in a fully connected manner to avoid noise. The temporal graph is a directed graph along the flow of time from t to $t+1$ and is represented as G_t^{time}.

$$G_{tij}^{\text{space}} = \frac{\exp \sigma_{tij}}{\sum_{j=1}^{N_t} \exp \sigma_{tij}} \tag{1.7}$$

where G_{tij}^{space} is the (i, j) th entry of G_t^{space} and is the amount of the spatial connectivity between the ith and jth objects at time step t and σ_{tij} denotes the IOU between the two objects.

$$G_{tij}^{\text{time}} = \frac{\exp\cos\left(o_t^i, o_{t+1}^j\right)}{\sum_{j=1}^{N_{t+1}} \exp\cos\left(o_t^i, o_{t+1}^j\right)} \tag{1.8}$$

where G_{tij}^{time} is the (i, j) th entry of G_t^{time} and $\cos\left(o^i, o^j\right)$ is the cosine similarity between the two feature vectors.

Both G_t^{space} and G_t^{time} are combined and sent to a graph convolutional network (GCN) for the updated object branch feature. Finally object branch feature is combined with scene branch feature and sent to the decoder for caption generation.

1.3.3 Comparison of results

We have shown the performance of different models (object detection and non-object detection based) on MSVD data. Microsoft video description (MSVD) dataset [30] consists of 1970 open domain videos. The videos are of single activity and with an average length of around 10–15 seconds. Each video is described by 40 captions generated by Amazon Mechanical Turk. To compare the performance, the four most popular metrics: CIDEr [31], METEOR [32], ROUGE-L [33], and BLEU-4 [34], have been mentioned. BLEU [34] calculates the geometric mean of modified precision between candidate and target n-gram tokens. METEOR [32] consider the synonyms during the calculation of the n-gram matching score. CIDEr [31] is originally

Table 1.2 Performance comparison on MSVD dataset. B@4, M, R, and C denote BLEU-4, METEOR, ROUGE_L, and CIDEr, respectively.

Models			MSVD			
	Detector	Type	B@4	M	R	C
SA-LSTM [22]	×	-	45.3	31.9	64.2	76.2
RecNet [23]	×	-	52.3	34.1	69.8	80.3
PickNet [24]	×	-	52.3	33.3	69.6	76.5
MARN [25]	×	-	48.6	35.1	71.9	92.2
GRU-EVE [26]	×	-	47.9	35.0	71.5	78.1
OA-BTG [27]	✓	Mask RCNN	**56.9**	36.2	-	90.6
STG-KD [29]	✓	Faster RCNN	52.2	**36.9**	73.9	93.0
ORG-TRL [28]	✓	Faster RCNN	54.3	36.4	73.9	95.2

designed to evaluate captioning tasks. From Table 1.2, we can see that methods with object features outperform non-object feature-based methods.

1.4 Conclusion

This chapter describes two main categories of object detection methods and their use cases. Two-stage region proposal-based methods are superior in performance metrics but not suitable for real-time applications. On the other hand, single-stage methods are fast but lack performance. We explain how object detection has helped improve video captioning, especially modeling the object-relational features over time. In video captioning, most object detection methods are based on two-stage proposal-based methods. Finally, the empirical results of video captioning show that object detection-based methods outperform others.

References

[1] Navneet Dalal and Bill Triggs. Histograms of oriented gradients for human detection. In *2005 IEEE Computer Society Conference on Computer Vision and Pattern Recognition (CVPR 2005), 20-26 June 2005, San Diego, CA, USA*, pages 886–893, 2005.

[2] David G. Lowe. Distinctive image features from scale-invariant keypoints. *Int. J. Comput. Vis.*, 60(2):91–110, 2004.

[3] Duc Thanh Nguyen, Zhimin Zong, Philip Ogunbona, and Wanqing Li. Object detection using non-redundant local binary patterns. In *Proceedings of the International Conference on Image Processing, ICIP 2010, September 26-29, Hong Kong, China*, pages 4609–4612, 2010.

[4] Corinna Cortes and Vladimir Vapnik. Support-vector networks. *Mach. Learn.*, 20(3):273–297, 1995.

[5] Leo Breiman. Random forests. *Mach. Learn.*, 45(1):5–32, 2001.

[6] Karen Simonyan and Andrew Zisserman. Very deep convolutional networks for large-scale image recognition. In Yoshua Bengio and Yann LeCun, editors, *3rd International Conference on Learning Representations, ICLR 2015, San Diego, CA, USA, May 7-9, 2015, Conference Track Proceedings*, 2015.

[7] Christian Szegedy, Sergey Ioffe, Vincent Vanhoucke, and Alexander A. Alemi. Inception-v4, inception-resnet and the impact of residual connections on learning. In Satinder P. Singh and Shaul Markovitch,

editors, *Proceedings of the Thirty-First AAAI Conference on Artificial Intelligence, February 4-9, 2017, San Francisco, California, USA*, pages 4278–4284, 2017.

[8] Kaiming He, Xiangyu Zhang, Shaoqing Ren, and Jian Sun. Deep residual learning for image recognition. In *2016 IEEE Conference on Computer Vision and Pattern Recognition, CVPR 2016, Las Vegas, NV, USA, June 27-30, 2016*, pages 770–778, 2016.

[9] Olga Russakovsky, Jia Deng, Hao Su, Jonathan Krause, Sanjeev Satheesh, Sean Ma, Zhiheng Huang, Andrej Karpathy, Aditya Khosla, Michael S. Bernstein, Alexander C. Berg, and Li Fei-Fei. Imagenet large scale visual recognition challenge. *Int. J. Comput. Vis.*, 115(3):211–252, 2015.

[10] Ross B. Girshick, Jeff Donahue, Trevor Darrell, and Jitendra Malik. Rich feature hierarchies for accurate object detection and semantic segmentation. In *2014 IEEE Conference on Computer Vision and Pattern Recognition, CVPR 2014, Columbus, OH, USA, June 23-28, 2014*, pages 580–587, 2014.

[11] Jasper R. R. Uijlings, Koen E. A. van de Sande, Theo Gevers, and Arnold W. M. Smeulders. Selective search for object recognition. *Int. J. Comput. Vis.*, 104(2):154–171, 2013.

[12] Kaiming He, Xiangyu Zhang, Shaoqing Ren, and Jian Sun. Spatial pyramid pooling in deep convolutional networks for visual recognition. *IEEE Trans. Pattern Anal. Mach. Intell.*, 37(9):1904–1916, 2015.

[13] Ross B. Girshick. Fast R-CNN. In *2015 IEEE International Conference on Computer Vision, ICCV 2015, Santiago, Chile, December 7-13, 2015*, pages 1440–1448, 2015.

[14] Shaoqing Ren, Kaiming He, Ross B. Girshick, and Jian Sun. Faster R-CNN: towards real-time object detection with region proposal networks. In Corinna Cortes, Neil D. Lawrence, Daniel D. Lee, Masashi Sugiyama, and Roman Garnett, editors, *Advances in Neural Information Processing Systems 28: Annual Conference on Neural Information Processing Systems 2015, December 7-12, 2015, Montreal, Quebec, Canada*, pages 91–99, 2015.

[15] Joseph Redmon, Santosh Kumar Divvala, Ross B. Girshick, and Ali Farhadi. You only look once: Unified, real-time object detection. In *2016 IEEE Conference on Computer Vision and Pattern Recognition, CVPR 2016, Las Vegas, NV, USA, June 27-30, 2016*, pages 779–788, 2016.

[16] Wei Liu, Dragomir Anguelov, Dumitru Erhan, Christian Szegedy, Scott E. Reed, Cheng-Yang Fu, and Alexander C. Berg. SSD: single

shot multibox detector. In Bastian Leibe, Jiri Matas, Nicu Sebe, and Max Welling, editors, *Computer Vision - ECCV 2016 - 14th European Conference, Amsterdam, The Netherlands, October 11-14, 2016, Proceedings, Part I*, volume 9905 of *Lecture Notes in Computer Science*, pages 21–37, 2016.

[17] Tsung-Yi Lin, Priya Goyal, Ross B. Girshick, Kaiming He, and Piotr Dollár. Focal loss for dense object detection. In *IEEE International Conference on Computer Vision, ICCV 2017, Venice, Italy, October 22-29, 2017*, pages 2999–3007, 2017.

[18] Kaiming He, Georgia Gkioxari, Piotr Dollár, and Ross B. Girshick. Mask R-CNN. In *IEEE International Conference on Computer Vision, ICCV 2017, Venice, Italy, October 22-29, 2017*, pages 2980–2988, 2017.

[19] Tsung-Yi Lin, Piotr Dollár, Ross B. Girshick, Kaiming He, Bharath Hariharan, and Serge J. Belongie. Feature pyramid networks for object detection. In *2017 IEEE Conference on Computer Vision and Pattern Recognition, CVPR 2017, Honolulu, HI, USA, July 21-26, 2017*, pages 936–944, 2017.

[20] Nicolas Carion, Francisco Massa, Gabriel Synnaeve, Nicolas Usunier, Alexander Kirillov, and Sergey Zagoruyko. End-to-end object detection with transformers. In Andrea Vedaldi, Horst Bischof, Thomas Brox, and Jan-Michael Frahm, editors, *Computer Vision - ECCV 2020 - 16th European Conference, Glasgow, UK, August 23-28, 2020, Proceedings, Part I*, volume 12346 of *Lecture Notes in Computer Science*, pages 213–229, 2020.

[21] Mark Everingham, Luc Van Gool, Christopher K. I. Williams, John M. Winn, and Andrew Zisserman. The pascal visual object classes (VOC) challenge. *Int. J. Comput. Vis.*, 88(2):303–338, 2010.

[22] Li Yao, Atousa Torabi, Kyunghyun Cho, Nicolas Ballas, Christopher J. Pal, Hugo Larochelle, and Aaron C. Courville. Describing videos by exploiting temporal structure. In *2015 IEEE International Conference on Computer Vision, ICCV 2015, Santiago, Chile, December 7-13, 2015*, pages 4507–4515, 2015.

[23] Bairui Wang, Lin Ma, Wei Zhang, and Wei Liu. Reconstruction network for video captioning. In *2018 IEEE Conference on Computer Vision and Pattern Recognition, CVPR 2018, Salt Lake City, UT, USA, June 18-22, 2018*, pages 7622–7631, 2018.

[24] Yangyu Chen, Shuhui Wang, Weigang Zhang, and Qingming Huang. Less is more: Picking informative frames for video captioning. In Vittorio Ferrari, Martial Hebert, Cristian Sminchisescu, and Yair Weiss,

editors, *Computer Vision - ECCV 2018 - 15th European Conference, Munich, Germany, September 8-14, 2018, Proceedings, Part XIII*, volume 11217 of *Lecture Notes in Computer Science*, pages 367–384, 2018.

[25] Wenjie Pei, Jiyuan Zhang, Xiangrong Wang, Lei Ke, Xiaoyong Shen, and Yu-Wing Tai. Memory-attended recurrent network for video captioning. In *IEEE Conference on Computer Vision and Pattern Recognition, CVPR 2019, Long Beach, CA, USA, June 16-20, 2019*, pages 8347–8356, 2019.

[26] Nayyer Aafaq, Naveed Akhtar, Wei Liu, Syed Zulqarnain Gilani, and Ajmal Mian. Spatio-temporal dynamics and semantic attribute enriched visual encoding for video captioning. In *IEEE Conference on Computer Vision and Pattern Recognition, CVPR 2019, Long Beach, CA, USA, June 16-20, 2019*, pages 12487–12496, 2019.

[27] Junchao Zhang and Yuxin Peng. Object-aware aggregation with bidirectional temporal graph for video captioning. In *IEEE Conference on Computer Vision and Pattern Recognition, CVPR 2019, Long Beach, CA, USA, June 16-20, 2019*, pages 8327–8336, 2019.

[28] Ziqi Zhang, Yaya Shi, Chunfeng Yuan, Bing Li, Peijin Wang, Weiming Hu, and Zheng-Jun Zha. Object relational graph with teacher-recommended learning for video captioning. In *2020 IEEE/CVF Conference on Computer Vision and Pattern Recognition, CVPR 2020, Seattle, WA, USA, June 13-19, 2020*, pages 13275–13285, 2020.

[29] Boxiao Pan, Haoye Cai, De-An Huang, Kuan-Hui Lee, Adrien Gaidon, Ehsan Adeli, and Juan Carlos Niebles. Spatio-temporal graph for video captioning with knowledge distillation. In *2020 IEEE/CVF Conference on Computer Vision and Pattern Recognition, CVPR 2020, Seattle, WA, USA, June 13-19, 2020*, pages 10867–10876, 2020.

[30] David L. Chen and William B. Dolan. Collecting highly parallel data for paraphrase evaluation. In Dekang Lin, Yuji Matsumoto, and Rada Mihalcea, editors, *The 49th Annual Meeting of the Association for Computational Linguistics: Human Language Technologies, Proceedings of the Conference, 19-24 June, 2011, Portland, Oregon, USA*, pages 190–200, 2011.

[31] Ramakrishna Vedantam, C. Lawrence Zitnick, and Devi Parikh. Cider: Consensus-based image description evaluation. In *IEEE Conference on Computer Vision and Pattern Recognition, CVPR 2015, Boston, MA, USA, June 7-12, 2015*, pages 4566–4575, 2015.

[32] Satanjeev Banerjee and Alon Lavie. METEOR: an automatic metric for MT evaluation with improved correlation with human judgments. In Jade Goldstein, Alon Lavie, Chin-Yew Lin, and Clare R. Voss, editors, *Proceedings of the Workshop on Intrinsic and Extrinsic Evaluation Measures for Machine Translation and/or Summarization@ACL 2005, Ann Arbor, Michigan, USA, June 29, 2005*, pages 65–72, 2005.

[33] Chin-Yew Lin. ROUGE: A package for automatic evaluation of summaries. In *Text Summarization Branches Out*, pages 74–81, Barcelona, Spain, July 2004. Association for Computational Linguistics.

[34] Kishore Papineni, Salim Roukos, Todd Ward, and Wei-Jing Zhu. Bleu: a method for automatic evaluation of machine translation. In *Proceedings of the 40th Annual Meeting of the Association for Computational Linguistics, July 6-12, 2002, Philadelphia, PA, USA*, pages 311–318, 2002.

2

A Deep Learning-based Framework for COVID-19 Identification using Chest X-Ray Images

Asifuzzaman Lasker[1], Mridul Ghosh[2], Sk Md Obaidullah[1], Chandan Chakraborty[3], and Kaushik Roy[4]

[1]Dept. of Computer Science & Engineering, Aliah University, India
[2]Department of Computer Science, Shyampur Siddheswari Mahavidyalaya, India
[3]Department of Computer Science & Engineering, NITTTR, India
[4]Department of Computer Science, West Bengal State University, India
E-mail: asifuzzaman.lasker@gmail.com; mridulxyz@gmail.com; sk.obaidullah@gmail.com; chakraborty.nitttrk@gmail.com; kaushik.mrg@gmail.com

Abstract

COVID-19 originally surfaced in Wuhan China quickly propagated over the globe and become a pandemic. This has had a disastrous consequence for people's regular life, healthcare, and the worldwide economy. It is indeed essential to find a positive individual as soon as feasible to avoid the pandemic from spreading wider and to cure infected individuals as promptly as necessary. A deep learning-based system was developed in this research to categorise four distinct kinds of chest X-ray pictures. We compared our result with the state-of-the-art architectures. It was observed that the proposed work gives us encouraging performance in the detection of COVID-19 among disparate sets of lung images captured normal as well as affected using homogeneous chest X-ray images.

Keywords: COVID-19, deep learning, cross-validation, chest X-ray.

2.1 Introduction

COVID-19 is a highly transmissible respiratory disease caused by the severe acute respiratory coronavirus 2 (SARS-CoV-2). SARS-CoV-2 is an RNA coronavirus responsible for the pandemic. The mutation rate of RNA viruses is substantially higher than that of their hosts, up to a million times higher. This characteristic has had a tremendous influence on world health and continues to be a major global issue [1], as the number of infected people and deaths continues to rise at an alarming rate.

People who have COVID-19 have a mild respiratory infection that can be treated without the use of drugs. People with medical issues such as diabetes, chronic respiratory disorders, and cardio-vascular diseases, on the other hand, are more susceptible to contract this virus. COVID-19 clinical signs are comparable with symptoms of the flu virus, involving temperature, exhaustion, chronic congestion, breathlessness, muscle cramps, and scratchy throat, by WHO statistics [2]. These frequent symptoms make early detection of the infection challenging. Because this is a virus, antibiotics that treat bacterial or fungal infections are unlikely to be effective. After the USA, India is the globe's second country in terms of number of individuals affected by illness aside from this position, it has a total of 36,582,129 cases with 485,350 deaths as on January 14, 2022.

Governments are advising people to test for COVID-19 through nasal swabs and (RT-PCR) lab report. The RT-PCR report, on the other hand, must be performed in a laboratory setting. Since such tests are time-consuming and expensive, they are not widely available in underdeveloped areas. Chest radiography like X-ray, CT, and MRI can help patients with respiratory disorders get a rapid diagnosis. Besides the RT-PCR test, chest scans reveal both the infection status (i.e., the existence or the lack of infection) and the severity of the disease. In addition, radiographic imaging is a quick and inexpensive technique.

These are very affordable equipment and maybe made quickly in isolated accommodation with the use of a transportable radiograph machine, reducing the risk of infection in hospitals. On the alternative, visual characteristics can be more challenging to identify, and it is much more difficult for unqualified experts. Researchers are trying to devise a novel approach to deal with a difficult situation. As a result, creating a computer-aided diagnosis system that can detect and classify lung diseases automatically will speed up the clinical diagnosis process [3]. The development of such a technology could help radiologists in the analysis of chest radiography (CXR) image data.

Given the present state of deep learning for medical images, applications such as segmentation and classification might be beneficial.

Artificial intelligence has progressed due to the tremendous increase in the machine and deep learning technique which are comparable to the human intelligence [4]. With the support of deep neural models, researchers were able to achieve excellent outcomes in object identification and tracking utilising computer vision techniques [5]. Deep learning algorithms have achieved significant results in image and video script recognition [6] since their development. Without using typical image processing methods and methodologies, the deep neural model processes the image or video data directly. The study of neural networks yielded previously inconceivable image prediction outcomes. For image classification, Convolutional Neural Network (ConvNet/CNN) is a well-known deep learning technique [7]. CNN had a lot of success with image recognition problems that directly from images, as well as mixed script identification [8]. CNN required less pre-processing steps and was able to quickly learn the fundamental properties.

The adoption of deep learning approaches acts as the supplemental of diagnostic devices for health professionals have sparked a lot of attention in current times. A deep learning-based system demonstrated reliable and effective diagnostic performance without human intervention [9]. The use of a deep neural systems could very well be beneficial for clinicians to identify COVID-19 accurately [10]. Using digital X-ray images, CAD systems effectively predicted many medical conditions including breast cancer [11], skin cancer [12, 13], and lung disorders [14]. The rapid spread of the this pandemic, which resulted in the death of millions of individuals from all over the globe, necessitates the use of deep learning technique to design solutions that are capable of improving identification efficiency. This is the key factor in the development of a a framework for identifying COVID-19 using complete digital images.

Here, a CNN-based architecture was developed to identify this disease among a different class of diseases from a set of X-ray images. Normal cases as well as for diseases like viral pneumonia and disease found in lung opacity-based images are identified by this architecture. The proposed CNN architecture is lightweight (i.e., the number of layers is less) keeping in mind to deploy in low resource platform. We also evaluate the presented approach's efficiency to the previously published work.

The cases (in million) and deaths (in thousand) of this diseases across different countries as reported by WHO [15] from year 2020 to 2022 is shown in Figure 2.1. This graph is based on WHO records from the beginning of

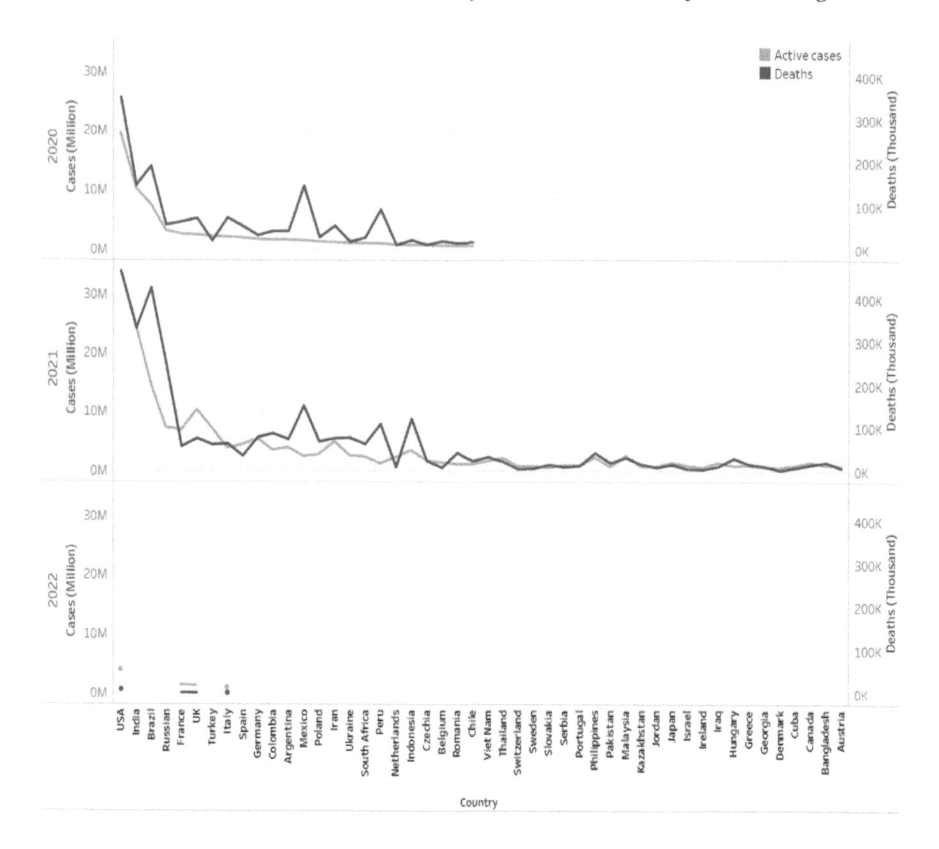

Figure 2.1 Cases and deaths, reported during the last three years (2020–2022) in different countries (> 600,000 cases).

2020 to January 1, 2022. The death rate and cases are shown country wise considering high to low in volume. Only those countries where the number of cases is more than 600,000 have been taken.

2.2 Related Works

Researchers from many disciplines along with policymakers urgently need to develop a strategy to control the unwanted situation. We explored some of the previously mentioned research areas on computer vision intelligent systems for automatic diagnosis and other graphic processing systems [16]. Researchers are using radiographic images to apply deep and machine learning algorithms as main categories to automatically classify COVID-19 and pneumonia disease. Ismael *et al.* [17] proposed methods using five pre-trained deep learning models for feature extraction. After extracting features, they used support vector machines (SVM) for classification. They used several linear kernel module such as quadradic, cubic, and gaussian. They also designed an end-to-end CNN model having 21 layers and a variety of parameters such as ReLU (rectified linear unit), batch normalization, pooling, and softmax. Singh *et al.* [18] utilized six public datasets with 15.465 samples in three classes (COVID-19, Pneumonia, and Normal). They increased the accuracy of the system by 6% when compared to the original sample by using two preprocessing techniques: image enhancement and image segmentation. To classify data of three separate classes, a stack-ensemble framework was used alongwith Naive Bayes as the meta learner. Similar to this architecture, a pre-trained model based on an ensemble approach was utilised by several other researchers to increase the performance of their architecture [19–21]. Other studies applied a variety of lightweight deep learning models with different architectures to improve prediction performance on test data even as limiting the number of trainable parameters. The best performing model was then used to the stack ensemble approach to improve reliability in additional low configuration devices [22].

Nayak *et al.* [23] used pre-trained models based on a public dataset and showed their accuracy of using ResNet-34 98.33%. An initial experiment of Varela-Santos *et al.* [24] used feature descriptors and CNN framework on their developed dataset of COVID-19 lung pictures. They considered as gray level co-occurrence matrix features (GLCM) and local binary patterns (LBP) as a texture feature extractor. Ahmed *et al.* [25] developed multi-level convolution filters in pre-processing that dynamically enhance the lung regions. DenseNet121 model was used in their augmented data. Authors in [26] developed a two-steps approaches to identify the COVID-19 problem. Two separate experiments were performed on a large and small dataset. In addition, they added two layers features extraction and feature smoothing. Habib *et al.* [27] introduced gene-based diagnosis technique. They also devised an ensemble method in their proposed model. Apart from this structural approach also designed CNN framework from x-ray images.

Paluru *et al.* [28] designed a lightweight CNN framework for resource constrained systems like Raspberry Pi 4, Android, and other devices. Their architecture was examined using three datasets and was put to the test under various experimental conditions. They tested their model with two different versions of the proposed architecture, one with low parameters and the other with high parameters. An EfficientNet model was used by Chowdhury *et al.* [29] and this model was fine-tuned in the top layers. Their classification performance was enhanced by applying soft and hard ensemble techniques. Furthermore, a mapping strategy is used to highlight areas that distinguish classes, allowing for a better overview of key understanding related to COVID-19. Paul *et al.* [30] used an ensemble of transfer-learning method to boost the efficiency of their deep learning systems for detecting COVID-19. In addition to implementing the ensemble technique, they used a reversed bell curve to provide weight to the classifiers. For training purposes, they used different datasets. The first dataset was a binary class which was taken from the IEEE COVID dataset, while the second data source was a ternary class derived from the Kaggle dataset. The accuracy achieved from first and second dataset was 99.66% and 99.84% respectively.

Oyelade *et al.* [31] pre-processed images by smoothening and denoising. Subsequently, to recognize COVID-19 situations, CNN model was used. Adam and other stochastic gradient descent (SGD) optimizers were used in this network. Monshi *et al.* [32] used VGG-19 and ResNet-50 for identifying COVID-19 on augmented data. Then they used EfficientNet-B0 with several augmentation methods and optimized CNN hyperparameters technique as a final experiment. A fusion-based model by using ensemble Monte Carlo (EMC) dropout and was proposed by Abdar *et al.* [33]. Their model is also a capable for handling noisy data. The adaptive zone growth approach was utilized by Alsaade *et al.* [34] to seperate out the bounding boxes from the lung X-ray radiographs after noise was removed from the pictures. Following segmentation, the features important for recognizing COVID-19 were extracted using a composite feature retrieval method combining 2D-DWT and GLCM. To extract the feature from the x-ray, Rezaee *et al.* [35] first employed three pre-trained transfer-learning models. In the feature selection stage, they used soft-voting using receiver operating characteristic (ROC), Entropy, and SNR methods, then for classification purposes they applied a SVM with RBF filter. Bharati *et al.* [36] combined VGG and spatial transform network. This three-pronged approach was then applied to a variety of parameters. They recommended this framework because basic CNN augmentation has poor performance in terms of rotation, tilted, and others.

Houssein *et al.* [37] implemented a hybrid method using random quantum circuits and CNN. This model can classify binary and multi-class cases. Another hybrid approach was followed in [38] by using CNN and discrete wavelet transforms (DWT). Before DWT and deep CNN were extracted features, the images were processed and segmented. With the least amount of duplication and the greatest amount of significance along with feature reduction, they were able to retrieve the best features. Lastly, random forest was exploited to complete the detection process. Khan *et al.* [39] developed hybrid feature extraction processed from multiple pre-trained deep learning models. They applied these techniques on two classes, three classes, and four class images, datasets containing 1341, 1341, and 449 images respectively.

2.3 Proposed Method

We selected chest X-ray images as the backbone of our experiment since it is sensitive to pose the differences among different diseases and is readily available in third-world countries as it is low cost as compared to other radiographic images. Also, it has been observed that the disease diagnosis with computer-aided methods which are being applied in radiography images have a significant false-positive rate [40].

Researchers have created a deep learning-based convolution neural network for diagnosing disease from radiography images [41] in order to reduce the high rate of false positives. Among different deep learning frameworks, CNN is a very popular technique [42] for processing spatial data. The design of such networks allows us to learn a wide range of complicated patterns that a basic neural network often fails to perform. CNN-based frameworks have a large number of applications including movie title extraction [43], self-driving cars, robots, surgical operations, etc. They extracted features from their CNN model and also separated the classes using the same network [44]. CNN can adapt effectively to train absolutely enormous medical data with numerous parameters, but a small sample size, that can lead to overfitting. It has lowered the total amount of variables that a basic model must learn in order to speed up learning in approximate anticipated rate. The pixel intensities from the source images are instantly sent to the image categorization feed forward net. Densely integrated systems are less effective than deep connectivity. Each built-in core matrices extracts features from the source as localized features to aid in imagery classification training.

In Figure 2.2, the output after applying a kernel having size 5×5 is convoluted with a color image (RGB) is shown. The relation with the input image size ($M\times M$), filter size ($f\times f$), stride (T) value, and padding (Z) is as follows:

$$(Mf + 2Z)/T + 1. \tag{2.1}$$

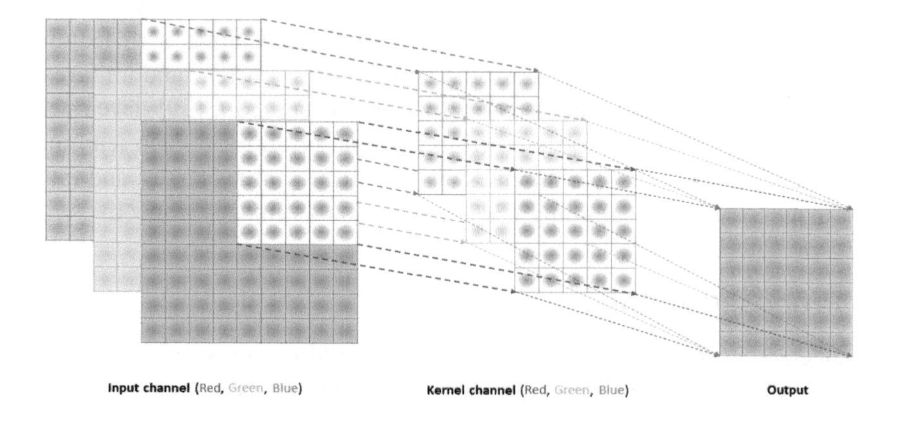

Figure 2.2 Three-color channel (Red, Green, Blue) with 5×5 kernel and output matrix.

According to Figure 2.2 the output feature dimension after convolutions becomes $(10 - 5 + 0)/1 + 1 = 6$, considering padding value as 0 and stride as 1.

Back propagation improves neural network performance by adjusting weight and bias value. The convolution layers use a variety of texture and sharpness features to detect COVID-19. The pooling layer helps to decrease the feature vector size coming out from the convolution layer. Multiple output matrices are generated for image categorization using multiple input channels. The output vector can be represented as shown in eqn 2.2. Here η denotes nonlinear in character, P, Q input and kernel matrix. X_i output matrix is computed as in eqn 2.2.

$$X_i = \eta(\sum_{j=1}^{n} P_j * Q_{j,i} + R_i). \tag{2.2}$$

With an application to the nonlinear activation function η, input matrix $P_{(j)}$ is convoluted with kernel $Q_{(j,i)}$. An output matrix X_i is generated by combining a number of convoluted matrices R_i with a bias.

Our proposed model is based on a CNN framework that includes triple convolution layers, triple max-pooling layers, and triple dense/fully connected layers. The proposed CNN framework was termed CPCPCPDDD where C denotes convolution, P denotes max pool, and D denotes dense layer.

There are 64, 32, and 32 numbers of filters having sizes (5×5), (5×5), and (3×3) used in Conv2d_1, Conv2d_2, and Conv2d_3, respectively. The max pool sizes are 3×3. A dropout value of 0.5 was considered in this framework. Each layer's activation function is ReLU, and the last dense layer's activation function is softmax.

The convolutional layer transfers the presence of features observed in individual portions of the input images into a feature vector. The process of creating a feature vector from this layer is as follows:

Where, * represent the convolution operator and $x^{r}_{j,k}$ denotes the filter in the $j\text{th}, k\text{th}$ element for $c\text{th}$ level on $\rho_{m,n}$ instance.

$$\nu_{m,n} = \sum_{m=1}^{m} \sum_{n=1}^{n} \rho_{(m-j,n-k)} * x^{r}_{j,k}. \tag{2.3}$$

Here, the ReLu activation function denotes as (λ)

$$\nu_{m,n} = \lambda(\sum_{m=1}^{m} \sum_{n=1}^{n} \rho_{(m-j,n-k)} * x^{r}_{j,k}) \tag{2.4}$$

$$\lambda(u) = \max(0; u). \tag{2.5}$$

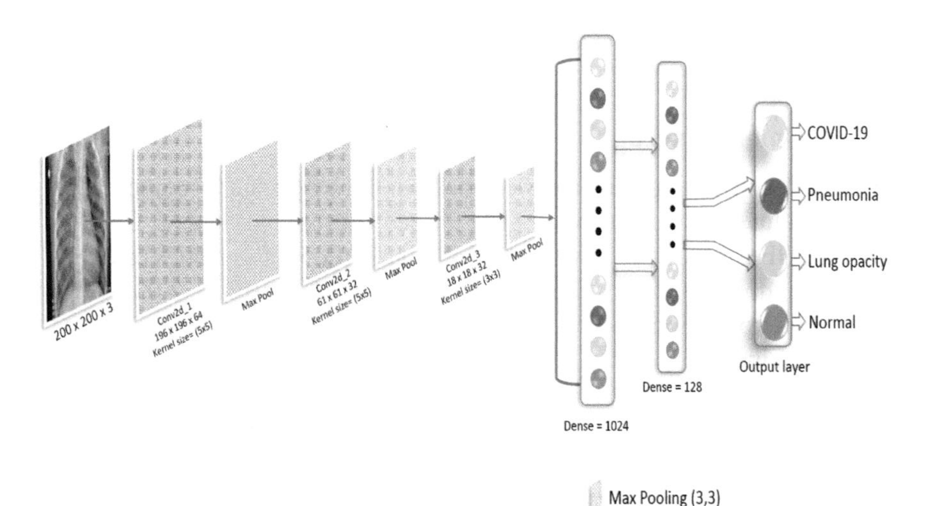

Figure 2.3 Proposed CPCPCPDDD architecture.

Table 2.1 The number of parameters generated for each layer of the proposed architecture.

Layers	Image dimension	#Parameter
conv2d_1 (Conv2D)	196, 196, 64	4864
max_pooling2d_1	65, 65, 64	0
conv2d_2 (Conv2D)	61, 61, 32	51,232
max_pooling2d_2	20, 20, 32	0
conv2d_3	18, 18, 32	9248
max_pooling2d_3	6, 6, 32	0
dropout_1	6, 6, 32	0
flatten_1	1152	0
dense_1	1024	1,180,672
dense_2	128	131,200
dense_3	4	516

Here, u represents neuron input [if ($u < 0$); Re(z) $= 0$ otherwise, Re(u) $= u$, where]. In the last dense layer, a softmax function is utilized, that can be formulated as

$$\chi(n) = \frac{e^n}{\sum_1^m e^n}. \tag{2.6}$$

Here, n signifies the source vector, which has a length of m. The proposed framework is shown in Figure 2.3 and the number of parameters generated corresponding to each layer is shown in Table 2.1.

2.4 Experiment

2.4.1 Experimental setup

We developed and tested our proposed methodology on a GPU computer featuring two Intel(R) Core (TM) i5-10400H CPUs running at 2.60 GHz, 2.59 Ghz, 32 GB of RAM, and two NVIDIA Quadro rtx 5000 graphics cards. It runs on Windows 10 Pro OS edition 20H2 using TensorFlow 2.0.0 for learning and interpretation, as well as cuDNN to expedite learning calculation on a GPU machine.

2.4.2 Ablation study

We performed an ablation study considering three parameters at different levels in the deep learning architecture, i.e., convolution kernel size, pooling size, and dense dimensions. In this study, we performed a total of 35 experiments starting with three layers as convolution, pooling, and dense. The performance obtained with a single layer in different types was unsatisfactory. As a result, we noticed that the addition of two dense layers (1024,3) improved the accuracy marginally. For the subsequent experiments, two convolution, two pooling, and two dense layers were added. Three experiments were conducted considering, different types of convolutional kernel pairs having dimensions (7×7, 5×5), (5×5, 3×3), and (5×5, 3×3) but the pooling size was the same i.e., (5×5, 3×3). In these two-layer type experiments, the dense size was altered in the third experiment (512, 3), which is shown in experiment number 7 in Table 2.2.

We have implemented seven different CNN structures so far, but the performance hasn't gained significantly. That's why our experiment added three layers in each type with different kernel sizes and dimensions. A total of 24 experiments were conducted in the three layers structure. In the whole experiment, we considered a different set of dense layers, in the first set the dense layers were kept (1024, 512, 3), second set it was (1024, 128, 3), in the third set (1024, 512, 3), and for the fourth set the dimensions consider as (256, 128, 3).

For experiment set-2 (i.e., experiment number 8 to 11) the convolution and pooling size were taken as (7×7, 7×7, 7×7) and (5×5, 5×5, 5×5), respectively. Similarly, from the experiment set-3 (i.e., experiment number 12 to 15) the convolution and pooling size were considered as (5×5, 5×5, 5×5) and (5×5, 5×5, 5×5). The convolution and pooling dimensions were kept as (5×5, 5×5, 5×5) and (3×3, 3×3, 3×3) in the experiment set-4 (i.e., experiment number 12 to 15). In the fifth, sixth, and seventh experimental sets, the convolution sets dimensions were kept as (3×3, 3×3, 3×3), (5×5, 3×3, 3×3) (5×5, 5×5, 3×3), respectively and pooling sets dimensions (3×3, 3×3, 3×3), (3×3,3×3,3×3), and (3×3, 3×3, 3×3), respectively.

It was observed that performance was decreased by more than 1% when convolution dimension change from (5×5, 5×5, 5×5) to (3×3, 3×3, 3×3) in experiment numbers 20 to 23. The performance is little improved when the first convolution size was altered from (3×3) to (5×5) which was realized from experiment numbers 24 to 27. Similarly, by changing the second convolution size from (3×3) to (5×5) the model accuracy marginally increased (reflected in exp. 28 to 31). So far, we conducted a total of 31 experiments, it was observed that dense dimensions (1024, 128, 3) produced higher accuracy than other dense combinations.

From these diverse types of experiments, it was observed that convolution kernel size 5×5, 5×5, 3×3, pooling size 3×3, 3×3, 3×3, and dense size 1024, 128, 3 provide the highest performance. In this architecture, we considered odd size kernel because once we apply even-sized kernels, there

Table 2.2 Ablation study based on convolution, pooling, and dense size.

Experiment No.	**Convolution size**	**Pooling**	**Dense**	**Accuracy (%)**
1	7×7	5×5	3	91.00
2	5×5	5×5	3	91.27
3	5×5	3×3	3	92.00
4	7×7, 5×5	5×5	1024, 3	92.15
5	5×5, 3×3	5×5, 3×3	1024, 3	92.35
6	5×5, 3×3	5×5, 3×3	1024, 3	92.45
7	5×5, 3×3	5×5, 3×3	512, 3	92.55
8	7×7, 7×7,7×7	5×5, 5×5, 5×5	1024, 128, 3	95.70
9	7×7, 7×7,7×7	5×5, 5×5, 5×5	1024, 512, 3	94.85
10	7×7, 7×7,7×7	5×5, 5×5, 5×5	512,256, 3	94.59
11	7×7, 7×7,7×7	5×5, 5×5, 5×5	256,128, 3	94.40
12	5×5, 5×5, 5×5	5×5, 5×5, 5×5	1024, 128, 3	96.92
13	5×5, 5×5, 5×5	5×5, 5×5, 5×5	1024, 512, 3	95.87
14	5×5, 5×5, 5×5	5×5, 5×5, 5×5	512,256, 3	95.75
15	5×5, 5×5, 5×5	5×5, 5×5, 5×5	256,128, 3	95.45
16	5×5, 5×5, 5×5	3×3, 3×3, 3×3	1024, 128, 3	97.45
17	5×5, 5×5, 5×5	3×3, 3×3, 3×3	1024, 512, 3	96.87
18	5×5, 5×5, 5×5	3×3, 3×3, 3×3	512,256, 3	96.65
19	5×5, 5×5, 5×5	3×3, 3×3, 3×3	256,128, 3	96.25
20	3×3, 3×3, 3×3	3×3, 3×3, 3×3	1024,128,3	96.72
21	3×3, 3×3, 3×3	3×3, 3×3, 3×3	1024, 512, 3	95.57
22	3×3, 3×3, 3×3	3×3, 3×3, 3×3	512,256, 3	95.35

Table 2.2 (Continued.)

Experiment No.	Convolution size	Pooling	Dense	Accuracy (%)
23	3×3, 3×3, 3×3	3×3, 3×3, 3×3	256,128, 3	95.15
24	5×5, 3×3, 3×3	3×3, 3×3, 3×3	1024,128,3	98.45
25	5×5, 3×3, 3×3	3×3, 3×3, 3×3	1024, 512, 3	98.27
26	5×5, 3×3, 3×3	3×3, 3×3, 3×3	512,256, 3	98.15
27	5×5, 3×3, 3×3	3×3, 3×3, 3×3	256,128, 3	97.95
28	5×5, 5×5, 3×3	3×3, 3×3, 3×3	1024, 128, 3	98.72
29	5×5, 5×5, 3×3	3×3, 3×3, 3×3	1024, 512, 3	98.57
30	5×5, 5×5, 3×3	3×3, 3×3, 3×3	512,256, 3	98.65
31	5×5, 5×5, 3×3	3×3, 3×3, 3×3	256,128, 3	98.35
32	5×5, 5×5, 3×3, 3×3	3×3, 3×3, 3×3	1024, 128, 3	98.68
33	5×5, 5×5, 3×3, 3×3	3×3, 3×3, 3×3	1024, 512, 3	98.47
34	5×5, 5×5, 3×3, 3×3	3×3, 3×3, 3×3	512, 256, 3	98.32
35	5×5, 5×5, 3×3, 3×3	3×3, 3×3, 3×3	256, 128, 3	98.15

will be distortions across the layers. In this case, filters with odd dimensions symmetrically split the pixels from the previous layer around the output pixel, and it is the principal reason for distortions.

2.4.3 Dataset

Researchers from various universities collaborated with physicians to prepare a dataset of chest X-ray images for COVID-19 positive [45], lung opacity, and normal and viral pneumonia. In lung opacity are non-COVID lung opacity class. The updated and released COVID-19 and different lung infection

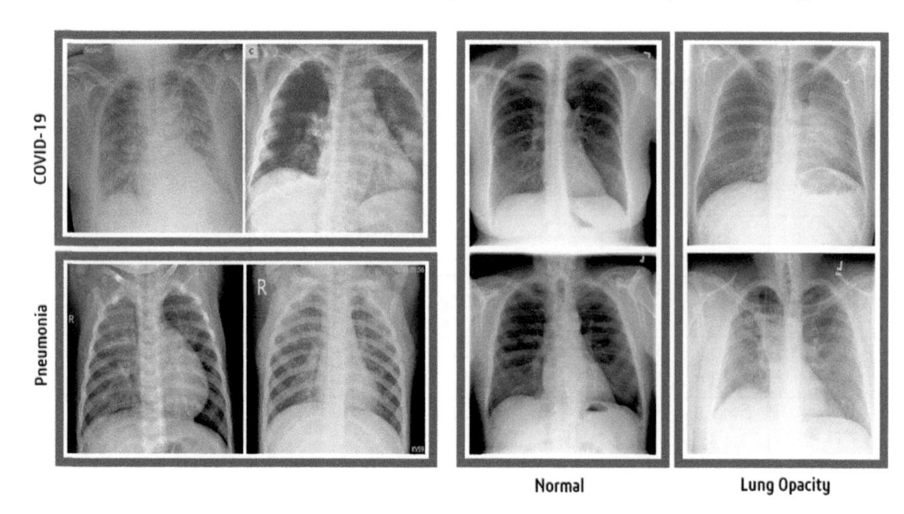

Figure 2.4 From left horizontal first block represents two COVID-19 X-ray images and second block shows pneumonia, vertical first block shows two normal X-ray images and second block represents lung opacity.

datasets are in phases. Figure 2.4 shows sample images from four classes. In the last available version, they release a total of 21,165 images, there are 3616 COVID-19, 1345 viral pneumonia, 6012 lung opacity, and 10,192 normal images. All images were in (.PNG) format, with dimensions 299×299. The Kaggle dataset was chosen because it is utilized in so many types of research throughout the world and allows for comparisons that might improve scientific research.

2.4.4 Evaluation protocol

A k-fold cross-validation approach was employed in this experiment. In this approach, the entire dataset is divided into k subsets, with one set serving as a testing set for verifying the framework and the other $(k-1)$ serving as a set of training sessions, and so on. Repeat this procedure to verify that all of the components have been trained and tested.

We employed a variety of evaluation measures to assess the efficacy of our deep learning-based cross-validation model. We experimented with different folding techniques and ran a large number of epochs to evaluate our system's performance. The following formula was used to compute the accuracy (Acc)

$$acc = \frac{\#correctly\_classified\_images}{\#Total\_images} * 100. \tag{2.7}$$

Since, accuracy alone is insufficient to demonstrate the system's performance, precision δ, sensitivity ε, and f-measure ω were also applied, which may be described as follows:

$$\delta = \frac{\gamma}{\gamma + \mu}, \varepsilon = \frac{\tau}{\tau + \varphi}, \omega = 2 * \frac{\delta * \varepsilon}{\delta + \varepsilon}. \tag{2.8}$$

Here γ, μ and φ denote true-positive, false-positive, and, false-negative rate, respectively.

Training regime: For script separation, the images were pre-processed and scaled to 200×200 pixels. The experiment performed five, seven, ten, and twelve-fold cross-validation. Initially, 100 epochs were used with a batch size of 100. A dropout value of 0.5 was utilized in this circumstance. The learning rate was calculated to be 0.001. For multi-class classification purposes, the Adam optimizer was used.

2.4.5 Result and analysis

Initially considering the batch size of 100 the experimentation was carried out by changing the cross-validation fold and evaluating the performance which is shown in Table 2.3.

For the five-fold cross-validation, 0.9694 F1 score and 96.98% accuracy were obtained. For seven-fold cross-validation, the accuracy improved by 1.08% and 10-fold cross-validation accuracy got improved by 0.49%. In 12-fold, 98.66% accuracy is obtained and if we increase the fold-number the accuracy falls down. Considering 12-fold cross-validation, a further experiment was performed by changing the batch size from 100 to 300 with a batch interval of 50. In Table 2.4, the results are depicted changing batch size. It is observed that for a batch size of 200, highest accuracy of 98.72% was obtained and increasing the batch size further the performance did not improve. Further experimentation was done considering the batch size of

Table 2.3 Evaluation matrices for various folds of cross-validation using batch size of 100.

Fold	Precision	Recall	F1 Score	Accuracy (%)
5-fold	0.9747	0.9645	0.9694	96.89
7-fold	0.9835	0.9760	0.9797	97.97
10-fold	0.9867	0.9842	0.9854	98.46
12-fold	0.9873	0.9872	0.9872	98.66
15-fold	0.9871	0.9870	0.9870	98.61

Table 2.4 The performance evaluation by changing the batch size from 100 to 300 with batch size interval 50 with 100 epoch and 12-fold cross cross-validation.

Batch Size	Precision	Recall	F1 Score	Accuracy (%)
100	0.9873	0.9872	0.9872	98.66
150	0.9872	0.9871	0.9873	98.69
200	**0.9873**	**0.9872**	**0.9875**	**98.72**
250	**0.9873**	**0.9872**	**0.9875**	**98.72**
300	0.9872	0.9871	0.9874	98.71

200 and 12-fold cross-validation, the number of epochs was increased from 100 to 500 with 100 epoch intervals. It was observed that the performance did not improve by changing the epoch. The same accuracy, precision, and recall of 98.72%, 0.9873, 0.9872, respectively were obtained for 100 to 500 epochs.

Further, we have evaluated the performance of different classes of COVID-19, Viral Pneumonia, and Normal considering batch size of 200, 100 epoch, and 12-fold cross-validation. An accuracy of 99.21%, and F1 Score of 0.9925 was achieved.

Error analysis

Using batch sizes of 200 and 100 epoch 12-fold cross-validation, the lowest error of 1.28% was obtained. From the confusion matrix in Figure 2.5, it is observed that out of 3616, 3573 images were correctly identified as COVID-19 and 23, 12, and 8 number of images were misclassified to lung opacity, normal, viral pneumonia classes, respectively. For lung opacity, 97.06% accuracy was obtained. There were 0.001% and 0.027% errors that occurred due to being wrongly classified as a COVID-19 and normal class. Out of 10,192 normal images 10,151 images were correctly classified as normal making misclassification of 8, 19, and 14 images as COVID-19, lung opacity, and viral pneumonia. It is seen that for viral pneumonia, the lowest error of 0.99% was obtained.

In Figure 2.6, few samples of misclassified images are presented. Here, image a, b, c, and d, are the original images of COVID-19, lung opacity, viral pneumonia, and normal, but they are misclassified as lung opacity, COVID-19, viral pneumonia, normal respectively. The possible reason for misclassification of lung opacity and COVID-19 is due to both types of images there is a large area of opaque space. For normal and viral pneumonia, misclassification occurred due to blurriness of noise.

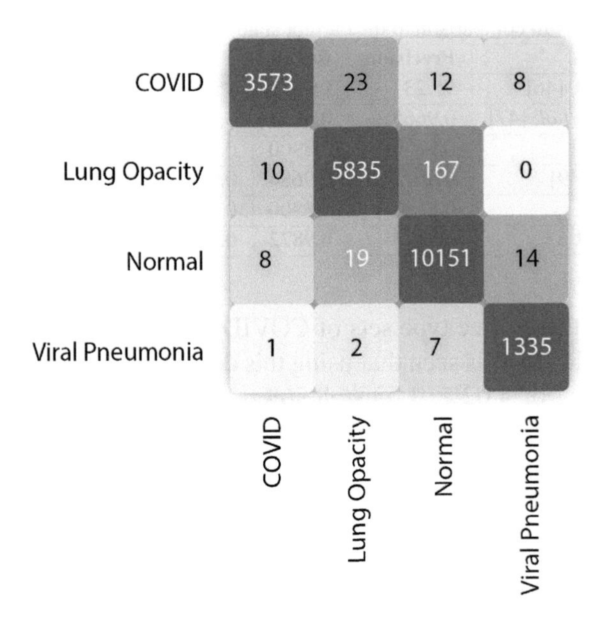

Figure 2.5 Confusion matrix of the highest result obtained after classification of four types of chest X-ray.

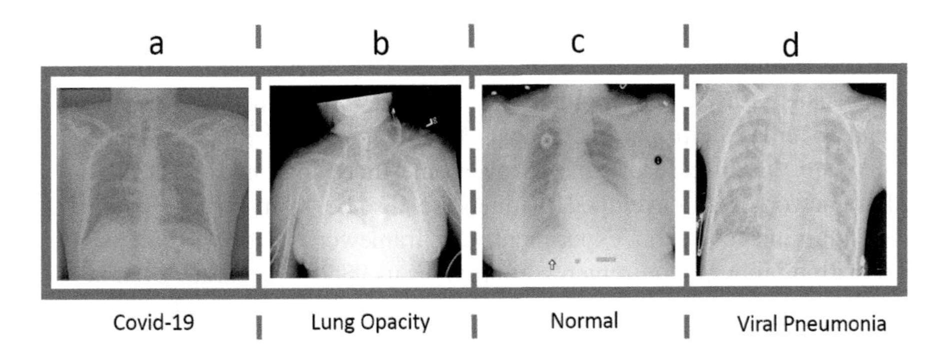

Figure 2.6 Representation of misclassification of images a, b, c, and d as lung opacity, COVID-19, viral pneumonia, normal respectively.

2.5 Comparison

We used precision, recall, F1 score, and accuracy measures to compare our results to the state-of-the-art approach, as shown in Table 2.5. Because the authors employed those datasets in their analyses, to make comparison fair

Table 2.5 In comparison to the current state-of-the-artÂămodel.

Authors	Precision	Recall	F1 Score	Accuracy (%)
Annavarapu *et al.* [46]	95.23	95.63	95.42	95.18
Chandrasekaran *et al.* [47]	0.9665	0.9654	0.9659	97.67
Ahmed *et al.* [48]	0.9200	0.8900	0.8980	90.64
Ouchicha *et al.* [49]	0.9672	0.9684	0.9668	96.69
Hertel *et al.* [50]	0.8600	0.8800	0.8583	90.66
Proposed model	**0.9873**	**0.9872**	**0.9875**	**99.21**

we also examined these three type sets of COVID-19, Viral pneumonia, and Normal. From Table 2.5, it is seen that using this developed framework, there are gain in accuracy of 4.02%, 1.53%, 8.56%, 2.51% compared with [46–49]. In comparison to existing deep learning models, our suggested model has a low false-positive rate, making it suitable for usage in real-world screening circumstances. Once the models have been trained, predictions for each case may be made in a matter of seconds, significantly lowering patient testing time.

2.6 Conclusion

In regions where COVID-19 testing kits are insufficient, detecting COVID-19 from chest X-ray images is essential for both physicians and patients to decrease diagnostic time and cost. In the current circumstance, parallel screening can be used to minimize infection from spreading to frontline staff and to create a primary diagnosis to identify whether a patient is infected with this disease or not. To overcome these limitations, a CNN-based framework was developed to diagnose disease from four different types of lung images in this experiment. Our framework is robust to detect disease with minimum requirements. It provides us 98.72% accuracy even in the poorest possible image situation. The use of our trustworthy technology can help to prevent infection spread. In the future, we'll create a system for COVID-19, Pneumonia, and SARS considering generative adversarial networks, attention-based encoder decoders, zero-shot learning, and other sophisticated deep learning architectures for better and reliable performance.

Conflict of Interest

The authors declare that there is no conflict of interest. The authors declare that there are no human participants and/or animals involved in this study.

The authors also declare that they have no known competing financial interests or personal relationships that could have appeared to influence the work reported in this paper.

References

[1] Himadri Mukherjee, Ankita Dhar, Sk Md Obaidullah, KC Santosh, and Kaushik Roy. Covid-19: A necessity for changes and innovations. In *COVID-19: Prediction, Decision-Making, and its Impacts*, pages 99–105. Springer, 2021.

[2] WHO. Coronavirus disease (COVID-19) is an infectious disease caused by the SARS-CoV-2 virus. https://www.who.int/health-topics/coronavirus#tab=tab_1/, 2019. [Online; accessed 2019].

[3] Sejuti Rahman, Sujan Sarker, Md Abdullah Al Miraj, Ragib Amin Nihal, AKM Nadimul Haque, and Abdullah Al Noman. Deep learning–driven automated detection of covid-19 from radiography images: a comparative analysis. *Cognitive Computation*, pages 1–30, 2021.

[4] Muhammad EH Chowdhury, Tawsifur Rahman, Amith Khandakar, Rashid Mazhar, Muhammad Abdul Kadir, Zaid Bin Mahbub, Khandakar Reajul Islam, Muhammad Salman Khan, Atif Iqbal, Nasser Al Emadi, et al. Can ai help in screening viral and covid-19 pneumonia? *IEEE Access*, 8:132665–132676, 2020.

[5] Ian Goodfellow, Yoshua Bengio, and Aaron Courville. *Deep learning*. MIT press, 2016.

[6] Mridul Ghosh, Himadri Mukherjee, Sk Md Obaidullah, KC Santosh, Nibaran Das, and Kaushik Roy. Lwsinet: A deep learning-based approach towards video script identification. *Multimedia Tools and Applications*, pages 1–34, 2021.

[7] Mercedes E Paoletti, Juan Mario Haut, Javier Plaza, and Antonio Plaza. A new deep convolutional neural network for fast hyperspectral image classification. *ISPRS journal of photogrammetry and remote sensing*, 145:120–147, 2018.

[8] Mridul Ghosh, Gourab Baidya, Himadri Mukherjee, Sk Md Obaidullah, and Kaushik Roy. A deep learning-based approach to single/mixed script-type identification. In *Advanced Computing and Systems for Security: Volume 13*, pages 121–132. Springer, 2022.

[9] Mugahed A Al-Antari, Mohammed A Al-Masni, Mun-Taek Choi, Seung-Moo Han, and Tae-Seong Kim. A fully integrated

computer-aided diagnosis system for digital x-ray mammograms via deep learning detection, segmentation, and classification. *International journal of medical informatics*, 117:44–54, 2018.

[10] Ahmad Alimadadi, Sachin Aryal, Ishan Manandhar, Patricia B Munroe, Bina Joe, and Xi Cheng. Artificial intelligence and machine learning to fight covid-19, 2020.

[11] Mohammed A Al-Masni, Mugahed A Al-Antari, Jeong-Min Park, Geon Gi, Tae-Yeon Kim, Patricio Rivera, Edwin Valarezo, Mun-Taek Choi, Seung-Moo Han, and Tae-Seong Kim. Simultaneous detection and classification of breast masses in digital mammograms via a deep learning yolo-based cad system. *Computer methods and programs in biomedicine*, 157:85–94, 2018.

[12] Sulaiman Vesal, Nishant Ravikumar, and Andreas Maier. Skinnet: A deep learning framework for skin lesion segmentation. In *2018 IEEE Nuclear Science Symposium and Medical Imaging Conference Proceedings (NSS/MIC)*, pages 1–3. IEEE, 2018.

[13] Mohammad H Jafari, Nader Karimi, Ebrahim Nasr-Esfahani, Shadrokh Samavi, S Mohamad R Soroushmehr, K Ward, and Kayvan Najarian. Skin lesion segmentation in clinical images using deep learning. In *2016 23rd International conference on pattern recognition (ICPR)*, pages 337–342. IEEE, 2016.

[14] Tulin Ozturk, Muhammed Talo, Eylul Azra Yildirim, Ulas Baran Baloglu, Ozal Yildirim, and U Rajendra Acharya. Automated detection of covid-19 cases using deep neural networks with x-ray images. *Computers in biology and medicine*, 121:103792, 2020.

[15] WHO. Daily cases and deaths by date reported to WHO. https://covid19.who.int/WHO-COVID-19-global-data.csv, 2022. [Online; accessed 2022].

[16] Mridul Ghosh, Sayan Saha Roy, Himadri Mukherjee, Sk Md Obaidullah, KC Santosh, and Kaushik Roy. Understanding movie poster: transfer-deep learning approach for graphic-rich text recognition. *The Visual Computer*, pages 1–20, 2021.

[17] Aras M Ismael and Abdulkadir Şengür. Deep learning approaches for covid-19 detection based on chest x-ray images. *Expert Systems with Applications*, 164:114054, 2021.

[18] Rajeev Kumar Singh, Rohan Pandey, and Rishie Nandhan Babu. Covidscreen: Explainable deep learning framework for differential diagnosis of covid-19 using chest x-rays. *Neural Computing and Applications*, pages 1–22, 2021.

[19] Kamini Upadhyay, Monika Agrawal, and Desh Deepak. Ensemble learning-based covid-19 detection by feature boosting in chest x-ray images. *IET Image Processing*, 14(16):4059–4066, 2020.

[20] Philip Meyer, Dominik Müller, Iñaki Soto-Rey, and Frank Kramer. Covid-19 image segmentation based on deep learning and ensemble learning. *Studies in Health Technology and Informatics*, 281:518–519, 2021.

[21] Xiaoshuo Li, Wenjun Tan, Pan Liu, Qinghua Zhou, and Jinzhu Yang. Classification of covid-19 chest ct images based on ensemble deep learning. *Journal of Healthcare Engineering*, 2021, 2021.

[22] Sivaramakrishnan Rajaraman, Jenifer Siegelman, Philip O Alderson, Lucas S Folio, Les R Folio, and Sameer K Antani. Iteratively pruned deep learning ensembles for covid-19 detection in chest x-rays. *IEEE Access*, 8:115041–115050, 2020.

[23] Soumya Ranjan Nayak, Deepak Ranjan Nayak, Utkarsh Sinha, Vaibhav Arora, and Ram Bilas Pachori. Application of deep learning techniques for detection of covid-19 cases using chest x-ray images: A comprehensive study. *Biomedical Signal Processing and Control*, 64:102365, 2021.

[24] Sergio Varela-Santos and Patricia Melin. A new approach for classifying coronavirus covid-19 based on its manifestation on chest x-rays using texture features and neural networks. *Information sciences*, 545:403–414, 2021.

[25] Sabbir Ahmed, Moi Hoon Yap, Maxine Tan, and Md Kamrul Hasan. Reconet: Multi-level preprocessing of chest x-rays for covid-19 detection using convolutional neural networks. *medRxiv*, 2020.

[26] Ruochi Zhang, Zhehao Guo, Yue Sun, Qi Lu, Zijian Xu, Zhaomin Yao, Meiyu Duan, Shuai Liu, Yanjiao Ren, Lan Huang, et al. Covid19xraynet: a two-step transfer learning model for the covid-19 detecting problem based on a limited number of chest x-ray images. *Interdisciplinary Sciences: Computational Life Sciences*, 12(4):555–565, 2020.

[27] Nahida Habib and Mohammad Motiur Rahman. Diagnosis of corona diseases from associated genes and x-ray images using machine learning algorithms and deep cnn. *Informatics in Medicine Unlocked*, page 100621, 2021.

[28] Naveen Paluru, Aveen Dayal, Håvard Bjørke Jenssen, Tomas Sakinis, Linga Reddy Cenkeramaddi, Jaya Prakash, and Phaneendra K Yalavarthy. Anam-net: Anamorphic depth embedding-based lightweight

cnn for segmentation of anomalies in covid-19 chest ct images. *IEEE Transactions on Neural Networks and Learning Systems*, 32(3):932–946, 2021.

[29] Nihad Karim Chowdhury, Muhammad Ashad Kabir, Md Muhtadir Rahman, and Noortaz Rezoana. Ecovnet: a highly effective ensemble based deep learning model for detecting covid-19. *PeerJ Computer Science*, 7:e551, 2021.

[30] Ashis Paul, Arpan Basu, Mufti Mahmud, M Shamim Kaiser, and Ram Sarkar. Inverted bell-curve-based ensemble of deep learning models for detection of covid-19 from chest x-rays. *Neural Computing and Applications*, pages 1–15, 2022.

[31] Olaide N Oyelade, Absalom E Ezugwu, and Haruna Chiroma. Covframenet: An enhanced deep learning framework for covid-19 detection. *IEEE Access*, 2021.

[32] Maram Mahmoud A Monshi, Josiah Poon, Vera Chung, and Fahad Mahmoud Monshi. Covidxraynet: Optimizing data augmentation and cnn hyperparameters for improved covid-19 detection from cxr. *Computers in biology and medicine*, 133:104375, 2021.

[33] Moloud Abdar, Soorena Salari, Sina Qahremani, Hak-Keung Lam, Fakhri Karray, Sadiq Hussain, Abbas Khosravi, U Rajendra Acharya, and Saeid Nahavandi. Uncertaintyfusenet: Robust uncertainty-aware hierarchical feature fusion with ensemble monte carlo dropout for covid-19 detection. *arXiv preprint arXiv:2105.08590*, 2021.

[34] Fawaz Waselallah Alsaade, HH Theyazn, and Mosleh Hmoud Al-Adhaileh. Developing a recognition system for classifying covid-19 using a convolutional neural network algorithm. *Computers, Materials, & Continua*, pages 805–819, 2021.

[35] Khosro Rezaee, Afsoon Badiei, and Saeed Meshgini. A hybrid deep transfer learning based approach for covid-19 classification in chest x-ray images. In *2020 27th National and 5th International Iranian Conference on Biomedical Engineering (ICBME)*, pages 234–241. IEEE, 2020.

[36] Subrato Bharati, Prajoy Podder, and M Rubaiyat Hossain Mondal. Hybrid deep learning for detecting lung diseases from x-ray images. *Informatics in Medicine Unlocked*, 20:100391, 2020.

[37] Essam H Houssein, Zainab Abohashima, Mohamed Elhoseny, and Waleed M Mohamed. Hybrid quantum convolutional neural networks model for covid-19 prediction using chest x-ray images. *arXiv preprint arXiv:2102.06535*, 2021.

[38] Rafid Mostafiz, Mohammad Shorif Uddin, Md Mahfuz Reza, Mohammad Motiur Rahman, et al. Covid-19 detection in chest x-ray through random forest classifier using a hybridization of deep cnn and dwt optimized features. *Journal of King Saud University-Computer and Information Sciences*, 2020.

[39] Md Saikat Islam Khan, Anichur Rahman, Md Razaul Karim, Nasima Islam Bithi, Shahab Band, Abdollah Dehzangi, and Hamid Alinejad-Rokny. Covidmulti-net: A parallel-dilated multi scale feature fusion architecture for the identification of covid-19 cases from chest x-ray images. *medRxiv*, 2021.

[40] Soumyaranjan Das. Can Artificial Intelligence is The Future of Breast Cancer Detection. https://myblogtuna.blogspot.com/2021/07/can-artificial-intelligence-is-future.html/, 2021. [Online; accessed 2021].

[41] Himadri Mukherjee, Subhankar Ghosh, Ankita Dhar, Sk Md Obaidullah, KC Santosh, and Kaushik Roy. Deep neural network to detect covid-19: one architecture for both ct scans and chest x-rays. *Applied Intelligence*, 51(5):2777–2789, 2021.

[42] Mridul Ghosh, Himadri Mukherjee, Sk Md Obaidullah, and Kaushik Roy. Stdnet: A cnn-based approach to single-/mixed-script detection. *Innovations in Systems and Software Engineering*, pages 1–12, 2021.

[43] Mridul Ghosh, Sayan Saha Roy, Himadri Mukherjee, Sk Md Obaidullah, Xiao-Zhi Gao, and Kaushik Roy. Movie title extraction and script separation using shallow convolution neural network. *IEEE Access*, 9:125184–125201, 2021.

[44] Himadri Mukherjee, Subhankar Ghosh, Ankita Dhar, Sk Md Obaidullah, KC Santosh, and Kaushik Roy. Shallow convolutional neural network for covid-19 outbreak screening using chest x-rays. *Cognitive Computation*, pages 1–14, 2021.

[45] Tawsifur Rahman, Amith Khandakar, Yazan Qiblawey, Anas Tahir, Serkan Kiranyaz, Saad Bin Abul Kashem, Mohammad Tariqul Islam, Somaya Al Maadeed, Susu M Zughaier, Muhammad Salman Khan, et al. Exploring the effect of image enhancement techniques on covid-19 detection using chest x-ray images. *Computers in biology and medicine*, 132:104319, 2021.

[46] Chandra Sekhara Rao Annavarapu et al. Deep learning-based improved snapshot ensemble technique for covid-19 chest x-ray classification. *Applied Intelligence*, 51(5):3104–3120, 2021.

[47] Ayan Kumar Das, Sidra Kalam, Chiranjeev Kumar, and Ditipriya Sinha. Tlcov-an automated covid-19 screening model using transfer learning

from chest x-ray images. *Chaos, Solitons & Fractals*, 144:110713, 2021.

[48] Faizan Ahmed, Syed Ahmad Chan Bukhari, and Fazel Keshtkar. A deep learning approach for covid-19 8 viral pneumonia screening with x-ray images. *Digital Government: Research and Practice*, 2(2):1–12, 2021.

[49] Chaimae Ouchicha, Ouafae Ammor, and Mohammed Meknassi. Cvdnet: A novel deep learning architecture for detection of coronavirus (covid-19) from chest x-ray images. *Chaos, Solitons & Fractals*, 140:110245, 2020.

[50] Robert Hertel and Rachid Benlamri. Cov-snet: A deep learning model for x-ray-based covid-19 classification. *Informatics in Medicine Unlocked*, page 100620, 2021.

3

Faster Region-based Convolutional Neural Networks for the Detection of Surface Defects in Aluminium Tubes

Vipul Sharma[1], Roohie Naaz Mir[1], and Mohammad Nayeem Teli[2]

[1]Department of Computer Science & Engineering, National Institute of Technology Srinagar, India
[2]Department of Computer Science, University of Maryland, College Park, United States
E-mail: vipul_1phd17@nitsri.net; naaz310@nitsri.net; nayeem@cs.umd.edu

Abstract

For high-quality products, surface fault identification is crucial. Surface defect identification on circular tubes, on the other hand, is more difficult than on flat plates due to the fact that the surface of circular tubes reflects light, resulting in overlooked faults. Surface defects on circular aluminum tubes, such as dents, bulges, foreign matter insertions, scratches, and cracks, were recognized using a unique faster region-based convolutional neural network (Faster RCNN) technique in this study. The proposed faster RCNN outperformed RCNN in terms of recognition speed and accuracy. Additionally, incorporating image enhancement into the approach improved recognition accuracy even more.

Keywords: RCNN, faster RCNN, surface defect detection, image recognition, image enhancement.

3.1 Introduction

Surface defect identification has long been regarded as a critical component of manufacturing. Defect identification on circular tubes, on the other hand, is more difficult than on flat plates due to the complicated structure of circular tubes. Liu *et al.* [1] suggested a modified multi-scale block local binary mode (LBP) technique, in which the picture is partitioned into small blocks and the image's eigenvector is a gray histogram. Because the size of the blocks is modified to find an acceptable scale to express fault characteristics, this algorithm is not only simple and efficient, but it also ensures excellent identification accuracy. Experiments proved the multi-scale block LBP algorithm's utility in online real-time detection systems. Peng *et al.* [2] suggested a method for detecting surface defects in cabinets based on irregularities in image moment characteristics, and established a gaussian distribution model for normal, defect-free image blocks. Anomaly features in defected image blocks were captured, and the defected image blocks were identified using a gaussian distribution model and a segmentation threshold. To reduce average run time and reinforce accuracy, Wang *et al.* [3] suggested an approach that incorporated an enhanced ResNet-50 with a reinforced faster region-based convolutional neural network (Faster RCNN). For fault segmentation and location, Tao *et al.* [4] devised a novel cascaded automatic encoder structure. Semantic segmentation was employed in the cascade network to turn the defected input images into pixel-level prediction templates. On the basis of low-order representations for texture, Zi *et al.* [5] proposed an unsupervised natural surface defect detection approach, in which the detection process was viewed as a new weighted low-order reconstruction model. Wu *et al.* [6] suggested a system calibration approach based on bias parameterization and realistic scanning trajectory modeling, as well as a constraint function for displaying image straightness and scale error. After that, the function was minimised to get the best estimate of system bias. This estimate was used to tweak the system as well as to reconstruct a reliable fault image. Lin *et al.* [7] developed a defect-enhanced generative adversarial network for detecting microcrack faults with high variety and apparent defect characteristics. Wei *et al.* [8] replaced the ROI pool with a weighted region of interest (ROI) pool, which corrected the regional misalignment induced by the two quantification methods. Han *et al.* [9] suggested a stacked convolutional auto encoder-based fault detection approach. On the basis of expert knowledge, the suggested auto encoder was trained with non-defective data and synthetic defect data produced with defect features. Zhao *et al.* [10] used a feature extraction network to reinforce identification for pointer surface flaws and recreated the convolutional layers for ResNet-50 with deformable convolution. Yang *et al.* [11] developed a tiny part defect detection system based on a single-short detection network combined with deep learning. Feng *et al.* [12] suggested a new orbital defect target detection technique. The algorithm's network architecture included a MobileNet backbone network and multiple unique detection layers with multi-scale feature mappings. In order to recognize carved characters on bearing dust covers, Lei *et al.* [13] used a character recognition method based on spatial pyramid character scale matching. Tian *et al.* [14] split components based on the plane light source image, then looked for faults in gray-level anomalies. Following that, the impact of reflection on the image was reduced by adjusting the camera exposure time based on the reflection characteristics of the part surface.

The position and direction of edges in the abnormal gray-level region of a multi-angle light source image assisted in determining whether or not an abnormal area was part of a defect.

Unlike the previous study, this one focuses on defect detection in aluminum round tubes, which must contend with issues like reflection and surface bending [15] [16].

3.2 Detection of Surface Flaws in Aluminum Tubes

Dents, bulges, foreign substance insertions, scratches, and cracks were all evaluated as surface defects on aluminum tubes in this study. Dents are caused by external forces on the tube surface; bulges are caused by welding spots bonding to the tube surface; foreign matter insertions are caused by the embedding of other metal substances during processing; scratches are caused by sharp objects passing through the tube; and cracks are caused by tool cutting causing tube cracking. To photograph faults, we use an industrial camera. The camera has a resolution of 2 megapixels. The frame rate can reach 30 frames per second.

The processing time is up to 3 milliseconds. The focal length of the lens is 3–8 mm. As seen in Figure 3.1, each flaw is manually marked first. At first, mark the defect with a large box. At the second time, use a small box to mark the defect. An industrial camera was used to image these faults, and 20 photographs of each category were taken, as shown in Figures 3.2 and 3.3.

Because of its curved shape and tendency to reflect light, aluminum tubes can cause flaw detection mistakes. Aluminum tubes have a lower flaw recognition rate than aluminum plates when using industrial cameras with the same resolution.

To improve the recognition rate of faster RCNN, the surface defect detection approach was integrated with picture enhancement using faster

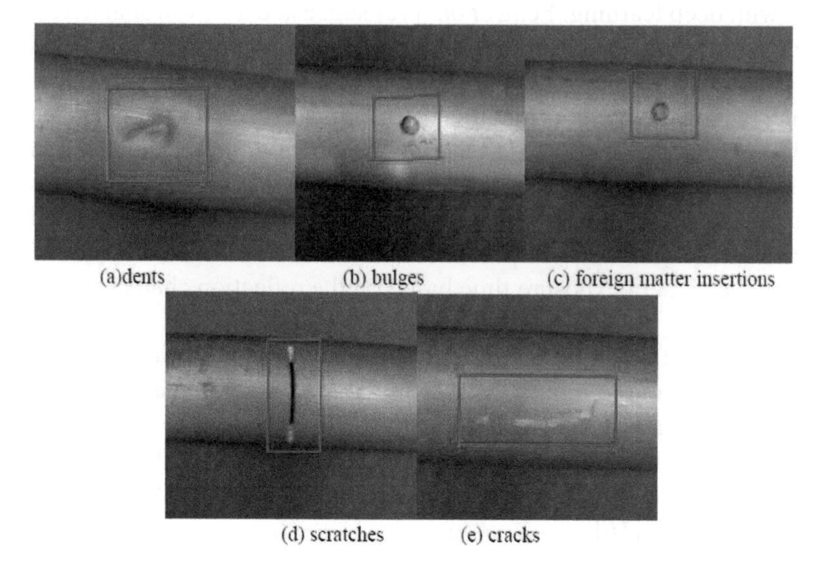

Figure 3.1 Surface defects in circular aluminum tubes.

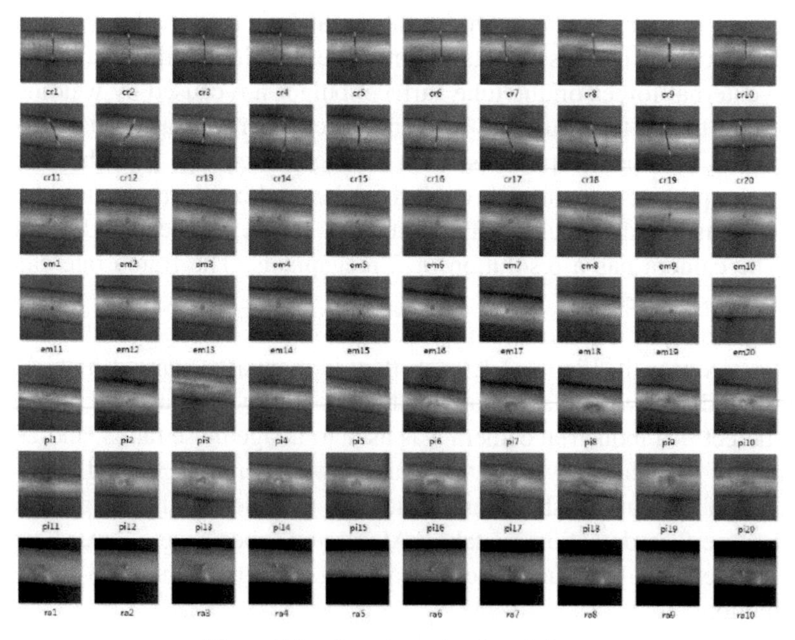

Figure 3.2 Overview of surface defects.

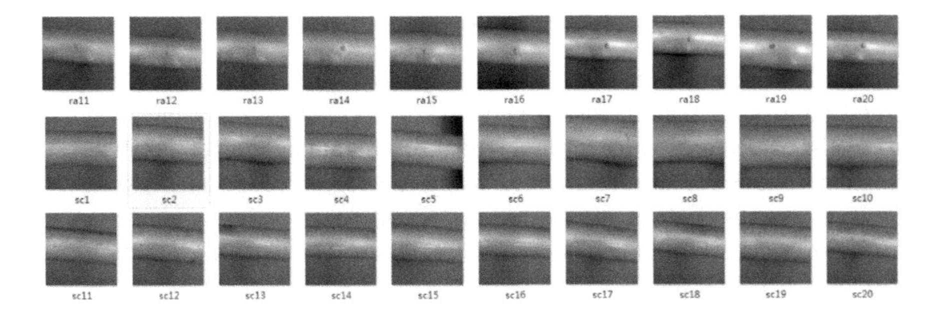

Figure 3.3 Some more samples depicting overview of surface defects.

RCNN in this work. Faster RCNN and RCNN recognition performances were compared.

3.3 RCNN-based Defect Recognition Method

RCNN involves the following four steps during detection:

1. The visual method, such as selective search, was used to generate numerous candidate regions.
2. Feature extraction was performed using CNN for each candidate region to form a high-dimensional feature vector.
3. These eigen values were inputted into a linear classifier to calculate the probability of a certain category for determining the contained objects.
4. A fine regression was performed on the location and size of the target peripheral box.

The network structure of RCNN is displayed in Figure 3.4.

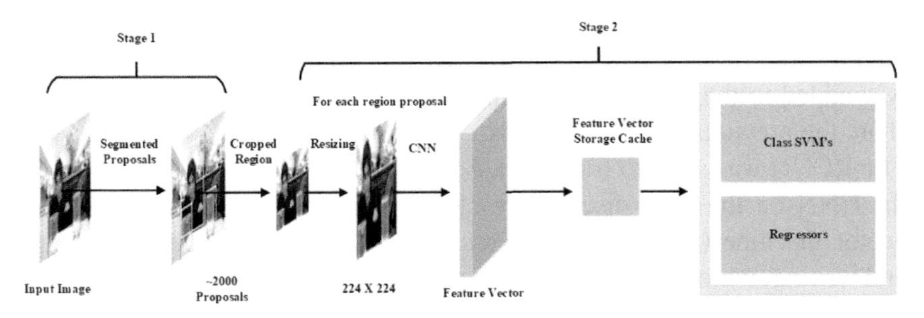

Figure 3.4 R-CNN architecture.

3.4 Faster RCNN-based Defect Recognition Method

Faster RCNN utilized feature maps created by CNNs at each stage, and the presented multi-level feature fusion network (MFN) was used to integrate many levels of features into a single feature containing defect location information. The ROIs were generated using the region proposal network (RPN), and the final detection results were produced using a detector consisting of a classifier and a border box regressor. The feature extraction network used was ResNet-50.

3.4.1 Improvements

The main differences between faster RCNN and RCNN are as follows:

1. An ROI pooling layer was introduced following the final convolutional layer.
2. The loss function adopted the multi-task loss function, which directly incorporated bounding box regression into the CNN network for training.
3. RPN replaced the former selective search method for generating the proposal window.
4. The CNN for generating the proposal window was shared with the CNN for target detection.

In comparison to RCNN, these differences allowed faster RCNN to achieve the following benefits.

1. In testing, RCNN was slow since it dissected an image into many proposal boxes. The image was created by stretching the proposed box, and each feature was extracted individually using CNN. There were many overlaps between these proposed boxes, and their eigen values might be shared. As a result, a significant amount of computer power was lost as a result of this event. The image was normalized and put directly into the CNN by the faster RCNN. Proposal box information was added to the feature map output by the last convolutional layer to share the previous CNN operations.
2. The RCNN was slow to train because it saved the features retrieved by the CNN on the hard disc before classifying them using the support vector machine (SVM). This operation required a lot of data reading and writing, which slowed down the training process. Faster RCNN, on the other hand, only required one image to be sent into the network. CNN features and proposal regions were extracted simultaneously for

each image. The training data were transmitted directly into the loss layer in graphics processing unit (GPU) memory, eliminating the need for repetitive calculation of the candidate area's first layers of features and huge data storage on the hard disc.

3. RCNN training necessitated a considerable amount of area. Many characteristics, such as training samples, were required for independent SVM classifiers and regressors, necessitating a huge amount of hard disc space. Faster RCNN used a deep network to do unified category judgment and location regression, which required no additional storage.

The network structure of faster RCNN is displayed in Figure 3.5.

Faster RCNN involves the following steps during detection.

1. Feature extraction: The CNN lowered network model complexity and weight number for extracting picture features. Images were used as network inputs in a straightforward manner. The VGG16 network was used to extract features. VGG16 outperformed LeNet, AlexNet, and ZFNet in terms of feature extraction depth and prominence, as well as detection performance.
2. RPN: RPN was used to generate high-quality region proposal boxes with image convolutional features similar to those of the detection network. When compared to the selective search strategy, the target detection

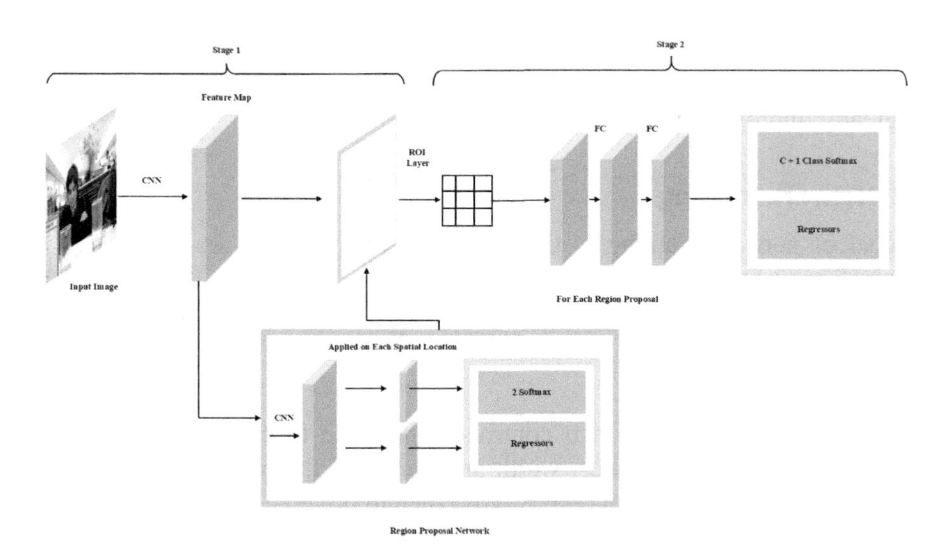

Figure 3.5 Faster RCNN architecture.

speed was significantly improved. A sliding window (3×3 convolution kernel) was used to slide across a convolution feature map obtained from the VGG16 network's first five convolutional layers (save for the max pool layer and the following fully connected layers). The sliding (convolution) process yielded a 512-dimensional vector, which was then fed into two parallel fully connected layers, the box-classification layer (cls) and the location regression layer (reg), to obtain classification and location information. Each sliding window's center corresponded to k anchors. Each anchor was associated with a specific dimension and aspect ratio. Three types of measurements and aspect ratios were used in RPN, resulting in nine anchors for each sliding window. Three sizes of boxes, large, medium, and small, were used for each point, with a ratio of 2:1, 1:2, and 1:1, respectively. As a result, nine area proposals were predicted simultaneously at each sliding window, resulting in 409 outputs from the box regression layer, which represented the coordinate coding of nine region proposal boxes. The likelihood that each proposal box was a target/nontarget was represented by the outputs of the box classification layer, which had 29 outputs.

3. Target recognition: After producing the proposal region with RPN, a faster RCNN was employed for detection and classification. With faster RCNN, end-to-end joint training was implemented with outstanding detection results. The convolution features extracted from the first five parts of VGG16 were shared by RPN and faster RCNN. The high-quality proposal region of RPN was used by faster RCNN in target recognition, greatly improving target detection speed.

3.4.2 Network training

The goal of network training is to reduce the loss function as much as possible. Eqn (3.1) gives the loss function.

$$F = \frac{1}{N_{cls}} \sum_i F_c\left(p_i, p_i^*\right) + a\frac{1}{N_{reg}} \sum_i F_r\left(c_i, c_i^*\right). \tag{3.1}$$

Here, a is the weight, which controls importance to classification or regression. C_i is the coordinates, width and height of the box. pi is equal to 1 for object; when it's not an object, p_i is equal to 0.

In network training, alternate RPN and faster RCNN were used.

RPN has to be educated first. The network was not trained with ImageNet but rather with the training data directly. When RPN training was completed, Model M1 was obtained, and it was utilized to construct proposal region P1.

Second, using the proposal area P1 generated by the RPN in the previous phase, faster RCNN was trained to yield model M2. M2 was also utilized to start RPN training and obtain model M3. Unlike the first stage, this phase of training focused on fine-tuning RPN parameters while fixing the shared convolutional layer parameters. Similarly, proposal region P2 was generated using the learned RPN. Lastly, M3 and P2 were used to train faster RCNN, which resulted in the final model M4. This phase of training only trained the fully connected layers of faster RCNN and calculated the parameters of the shared convolutional layers. As a result, these two networks had the identical convolutional layers and were combined to form a single network.

After deleting the bounding boxes from the original 20,000 boxes, there were only 6000 boxes left. There were 2000 boxes left after screening some that overlapped and resembled each other. The first 128 boxes were kept after sorting, reducing the number of boxes and improving detection speed.

3.5 Faster RCNN and Image Enhancement-based Defect Detection Method

Image preprocessing techniques could be used to improve image contrast and enhance image features. In this experiment, histogram equalization and filter smoothing were utilized as preprocessing techniques.

By linearly changing the gray level into a specific range, contrast enhancement boosted the contrast.

Equalization of contrast using histograms is referred to as histogram equalization.

$$s_k = \frac{L-1}{MN}\sum_{j=0}^{k} n_j, \quad k = 0, 1, 2, \ldots, L-1. \tag{3.2}$$

In eqn (3.2), MN is the total number of pixels in the image, n_j denotes the number of pixels with the r_j gray level, and L denotes the number of potential gray levels in the image (e.g., L = 256 for 8-bit images). The gray levels of the pixels in the output image could be derived by mapping n_j in the input image to s_k using the equation. T(r_k), the equation's transformation (mapping), denotes histogram equalization or histogram linear transformation [17] [18].

The median filtering approach, in which the median of the gray levels of points is used to replace the gray level of the center pixel in the vicinity of the smoothing template, can be represented as follows:

$$g(x,y) = \text{Med}\{f_i, f_{i+1}, \ldots, f_{i+v}\} = \begin{cases} f_{\frac{v+1}{2}} & \text{v is an odd number} \\ \frac{1}{2}\left(f_{\frac{v}{2}} + f_{\frac{v+1}{2}}\right) & \text{v is an even number}. \end{cases} \tag{3.3}$$

The gray level of the points in the area is represented by f_i in eqn (3.3) [19] [20].

The starting learning rate for training was set to 0.001, and each iteration used a quantity containing 32 observed values. The number of iterations was set to 100. RPN's effective maximum and minimum thresholds were set to 0.7 and 0.3, respectively. When establishing the category of the regression box, a maximum threshold of 0.5 and a minimum threshold of 0.1 were set.

The first time, mark the problem with a large box, and the second time, with a small box. Experiments have shown that the size of the box has a significant impact on the recognition rate.

3.6 Results

The recognition effect of the three algorithms was assessed using category average pixel accuracy and test set recognition accuracy. Faster RCNN outperformed RCNN in terms of recognition accuracy, while faster RCNN+preprocessing outperformed RCNN even more. The first labeled

Table 3.1 Defect identification results after the first marking.

Algorithm	Category average pixel accuracy	Test set recognition accuracy
RCNN	0.798	0.88
Faster RCNN	0.799	0.92
Faster RCNN +Preprocessing	0.85	0.96

Table 3.2 Defect identification results after the second marking.

Algorithm	Category average pixel accuracy	Test set recognition accuracy
RCNN	0.62	0.893
Faster RCNN	0.603	0.907
Faster RCNN +Preprocessing	0.61	0.915

Table 3.3 Comparative results with state-of-the art after first marking for defect identification.

Algorithm	Category average pixel accuracy	Test set recognition accuracy
R-CNN BB [21]	0.68	0.74
R-CNN VGG [21]	0.62	0.69
R-CNN VGG BB [21]	0.66	0.73
Fast R-CNN [22]	0.72	0.80
HyperNet VGG [23]	0.68	0.76
HyperNet SP [23]	0.64	0.79
Feature Edit [24]	0.72	0.83
SDS [25]	0.67	0.83
NoC [26]	0.52	0.67
MR_CNN S CNN [27]	0.56	0.65
MR CNN [27]	0.64	0.76
Fast R-CNN + YOLO [23]	0.72	0.85
NUS_NIN [23]	0.56	0.67
YOLO [23]	0.64	0.78
Baby Learning [28]	0.67	0.71
Faster RCNN +Preprocessing	0.85	0.96

Table 3.4 Comparative results with state-of-the art after first marking for defect identification.

Algorithm	Category average pixel accuracy	Test set recognition accuracy
R-CNN BB [21]	0.69	0.76
R-CNN VGG [21]	0.63	0.71
R-CNN VGG BB [21]	0.67	0.75
Fast R-CNN [22]	0.73	0.82
HyperNet VGG [23]	0.69	0.78
HyperNet SP [23]	0.65	0.81
Feature Edit [24]	0.73	0.85
SDS [25]	0.68	0.85
NoC [26]	0.53	0.69
MR_CNN S CNN [27]	0.57	0.67
MR CNN [27]	0.65	0.78
Fast R-CNN + YOLO [23]	0.73	0.87
NUS_NIN [23]	0.57	0.69
YOLO [23]	0.65	0.80
Baby Learning [28]	0.68	0.73
Faster RCNN +Preprocessing	0.87	0.96

image was sloppy and imprecise, and the model's performance could only be improved to a certain extent after training on this dataset. The findings of flaw detection are shown in Table 3.1. We re-annotated the data and achieved a nice result this time. As indicated in Table 3.2, the recognition accuracy for the aluminum tube was 96 %, indicating that the flaw detection was satisfactory.

In Tables 3.3 and 3.4, we have even compared faster RCNN + preprocessing technique with state-of-the-art for both first and second markings.

3.7 Conclusion

Because of the many reflections, identifying surface defects on circular aluminum tubes is more difficult than on flat plates. Five surface flaws, specifically aluminum tubes, were explored in this study, comprising dents, bulges, foreign matter insertions, scratches, and fractures. For defect recognition, a method combining faster RCNN and image enhancement was applied, with a recognition accuracy of 96 %, indicating good fault detection ability.The photographs that were first labeled were sloppy and inaccurate. As a result, after a given amount of training with this dataset, the model's performance could not be enhanced any more. The updated dataset produced satisfactory findings once the data were relabeled in detail.

References

[1] Yang Liu, Ke Xu, and Jinwu Xu. An improved mb-lbp defect recognition approach for the surface of steel plates. *Applied Sciences*, 9(20):4222, 2019.

[2] Yeping Peng, Songbo Ruan, Guangzhong Cao, Sudan Huang, Ngaiming Kwok, and Shengxi Zhou. Automated product boundary defect detection based on image moment feature anomaly. *IEEE Access*, 7:52731–52742, 2019.

[3] Shuai Wang, Xiaojun Xia, Lanqing Ye, and Binbin Yang. Automatic detection and classification of steel surface defect using deep convolutional neural networks. *Metals*, 11(3):388, 2021.

[4] Xian Tao, Dapeng Zhang, Wenzhi Ma, Xilong Liu, and De Xu. Automatic metallic surface defect detection and recognition with convolutional neural networks. *Applied Sciences*, 8(9):1575, 2018.

[5] Qizi Huangpeng, Hong Zhang, Xiangrong Zeng, and Wenwei Huang. Automatic visual defect detection using texture prior and low-rank representation. *IEEE Access*, 6:37965–37976, 2018.

[6] Fan Wu, Pin Cao, Yubin Du, Haotian Hu, and Yongying Yang. Calibration and image reconstruction in a spot scanning detection system for surface defects. *Applied Sciences*, 10(7):2503, 2020.

[7] Xiaoming Lv, Fajie Duan, Jia-jia Jiang, Xiao Fu, and Lin Gan. Deep metallic surface defect detection: The new benchmark and detection network. *Sensors*, 20(6):1562, 2020.

[8] Song Lin, Zhiyong He, and Lining Sun. Defect enhancement generative adversarial network for enlarging data set of microcrack defect. *IEEE Access*, 7:148413–148423, 2019.

[9] Manhuai Lu and Chin-Ling Chen. Detection and classification of bearing surface defects based on machine vision. *Applied Sciences*, 11(4):1825, 2021.

[10] Wenzhuo Zhang, Juan Hu, Guoxiong Zhou, and Mingfang He. Detection of apple defects based on the fcm-npga and a multivariate image analysis. *IEEE Access*, 8:38833–38845, 2020.

[11] Rubo Wei, Yonghong Song, and Yuanlin Zhang. Enhanced faster region convolutional neural networks for steel surface defect detection. *ISIJ International*, 60(3):539–545, 2020.

[12] Young-Joo Han and Ha-Jin Yu. Fabric defect detection system using stacked convolutional denoising auto-encoders trained with synthetic defect data. *Applied Sciences*, 10(7):2511, 2020.

[13] Weidong Zhao, Hancheng Huang, Dan Li, Feng Chen, and Wei Cheng. Pointer defect detection based on transfer learning and improved cascade-rcnn. *Sensors*, 20(17):4939, 2020.

[14] Jing Yang, Shaobo Li, Zheng Wang, and Guanci Yang. Real-time tiny part defect detection system in manufacturing using deep learning. *IEEE Access*, 7:89278–89291, 2019.

[15] Satoshi Yamaguchi, Eiichi Sato, Yasuyuki Oda, Ryuji Nakamura, Hirobumi Oikawa, Tomonori Yabuushi, Hisanori Ariga, and Shigeru Ehara. Zero-dark-counting high-speed x-ray photon detection using a cerium-doped yttrium aluminum perovskite crystal and a small photomultiplier tube and its application to gadolinium imaging. *Japanese Journal of Applied Physics*, 53(4):040304, 2014.

[16] Ning Lang, Decheng Wang, Peng Cheng, Shanchao Zuo, and Pengfei Zhang. Virtual-sample-based defect detection algorithm for aluminum tube surface. *Measurement Science and Technology*, 32(8):085001, 2021.

[17] Yu Wang, Qian Chen, and Baeomin Zhang. Image enhancement based on equal area dualistic sub-image histogram equalization method. *IEEE transactions on Consumer Electronics*, 45(1):68–75, 1999.

[18] Jameel Ahmed Bhutto, Tian Lianfang, Qiliang Du, Toufique Ahmed Soomro, Yu Lubin, and Muhammad Faizan Tahir. An enhanced image fusion algorithm by combined histogram equalization and fast gray level grouping using multi-scale decomposition and gray-pca. *IEEE Access*, 8:157005–157021, 2020.

[19] Bharat Garg and GK Sharma. A quality-aware energy-scalable gaussian smoothing filter for image processing applications. *Microprocessors and Microsystems*, 45:1–9, 2016.

[20] Yupei Yan and Yangmin Li. Mobile robot autonomous path planning based on fuzzy logic and filter smoothing in dynamic environment. In *2016 12th World congress on intelligent control and automation (WCICA)*, pages 1479–1484. IEEE, 2016.

[21] Ross Girshick, Jeff Donahue, Trevor Darrell, and Jitendra Malik. Rich feature hierarchies for accurate object detection and semantic segmentation. In *Proceedings of the IEEE conference on computer vision and pattern recognition*, pages 580–587, 2014.

[22] Ross Girshick. Fast r-cnn. In *Proceedings of the IEEE international conference on computer vision*, pages 1440–1448, 2015.

[23] Joseph Redmon, Santosh Divvala, Ross Girshick, and Ali Farhadi. You only look once: Unified, real-time object detection. In *Proceedings of the IEEE conference on computer vision and pattern recognition*, pages 779–788, 2016.

[24] Zhiqiang Shen and Xiangyang Xue. Do more dropouts in pool5 feature maps for better object detection. *arXiv preprint arXiv:1409.6911*, 2014.

[25] Bharath Hariharan, Pablo Arbeláez, Ross Girshick, and Jitendra Malik. Simultaneous detection and segmentation. In *European Conference on Computer Vision*, pages 297–312. Springer, 2014.

[26] Shaoqing Ren, Kaiming He, Ross Girshick, Xiangyu Zhang, and Jian Sun. Object detection networks on convolutional feature maps. *IEEE*

transactions on pattern analysis and machine intelligence, 39(7):1476–1481, 2017.

[27] Spyros Gidaris and Nikos Komodakis. Object detection via a multi-region and semantic segmentation-aware cnn model. In *Proceedings of the IEEE International Conference on Computer Vision*, pages 1134–1142, 2015.

[28] Jian Dong, Qiang Chen, Shuicheng Yan, and Alan Loddon Yuille. Towards unified object detection and semantic segmentation. In *ECCV*, 2014.

4

Real Time Face Detection-based Automobile Safety System using Computer Vision and Supervised Machine Learning

Navpreet Singh Kapoor[1], Mansimar Anand[1], Priyanshu[1], Shailendra Tiwari[1], Shivendra Shivani[1], and Raman Singh[2]

[1]Department of Computer Science & Engineering, Thapar Institute of Engineering and Technology, India
[2]School of Computing, Engineering and Physical Sciences, University of the West of Scotland, United Kingdom
E-mail: navpreetsingh181124@gmail.com; anandmansimar@gmail.com; priyanshuagg08@gmail.com; shailendra@thapar.edu; shivendra.shivani@thapar.edu; raman.singh@uws.ac.uk

Abstract

Improvements in vehicle safety measures are critical for addressing national and international reductions in road fatalities and producing a safer road traffic system. The safety of vehicles on roads directly concerns the safety of all people who use the roadways. It primarily consists of steps intended to help avoid or lessen the likelihood of injury in the case of a collision. Despite being a significant public health concern, road traffic injuries have been overlooked, necessitating a concerted effort to develop a long-term detection and prevention plan. Road transportation is one of the most complicated and dangerous public infrastructures and systems regularly available to the general public. According to the survey by World Health Organization, the estimated number of people killed in road traffic crashes each year is nearly 1.3 million, with an estimated number of people injured in such accidents ranging between 20 and 50 million. Road traffic crashes are the leading cause of death for children and young adults aged 5–29 years. Technology has advanced dramatically in nearly every sector since the dawn of the computer age. This chapter discusses the leading working methods of the leading automobile safety technology, offers a detection and prevention system, and describes its specific working methods.

Keywords: Road traffic accidents, accident prevention, drowsiness detection, machine learning, computer vision, face detection, automatic braking system.

4.1 Introduction

Currently, road transport systems are an essential part of human activities, and along with this essential part, there comes an issue of drowsiness or distraction, which is also a major concern. We all can be the victim of drowsiness while driving, simply after a too-short night's sleep, altered physical condition, or during long journeys [1]. The sensation of rest reduces the driver's level of vigilance, producing dangerous situations and increasing the probability of accidents. Driver drowsiness and fatigue are among the crucial causes of road accidents. Till we have focused on drowsiness, but along with this, there is one more condition that may cause accidents which is distraction. In simple terms, when you are not paying attention to the road you are driving on, you are said to be distracted. Using your phone while driving, falling asleep, looking somewhere else, etc., are the various reasons which may lead to distraction and ultimately to an accident. Road accidents are recently growing and have become a concern for the government and regulatory boards. India ranks first in the amount of deaths related to road accidents across the 199 countries and accounts for almost 11% of the accident-related deaths in the World. In 2021, India's Union Minister Mr. Nitin Gadkari cited "Road accidents are more dangerous than the COVID-19 pandemic," reporting a loss of 1.50 lakh of lives compared to 1.46 lakh because of COVID-19. This translates to a loss of INR 91.16 lakhs per person lost, which accounts for 3.14 % of GDP annually [2].

The data shows a direct correlation between the number of automobiles sales and road accidents a year [3]. As the number of sales increases, there is an increase in the number of accidents, as shown in Figures 4.1 and 4.2. In this context, it is important to use new technologies to design and build systems that can monitor drivers and measure their level of attention during the entire driving process. In response, it should be capable of alerting the driver or at least provide some signals to gather the driver's attention back. If necessary, activating the automatic braking system.

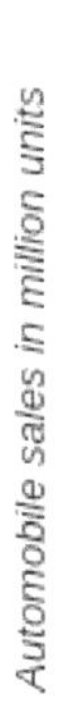

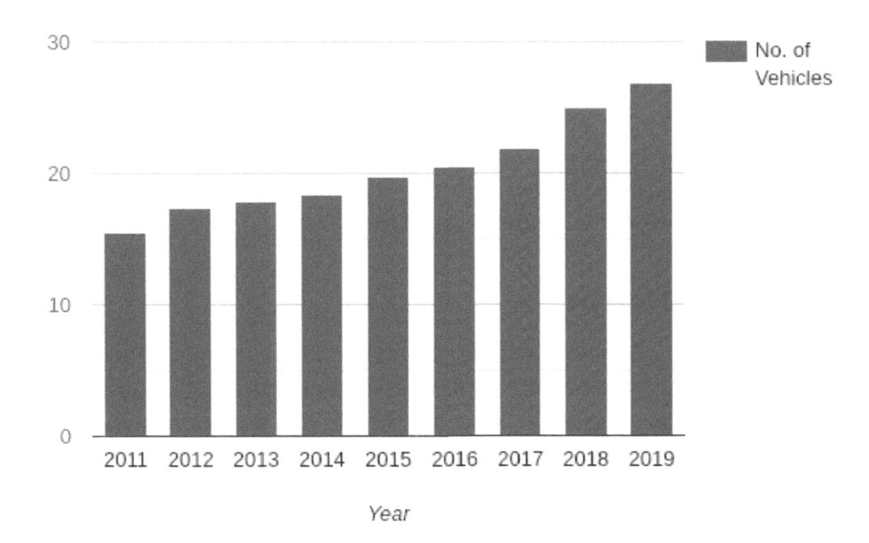

Figure 4.1 The number of automobiles sales in millions.

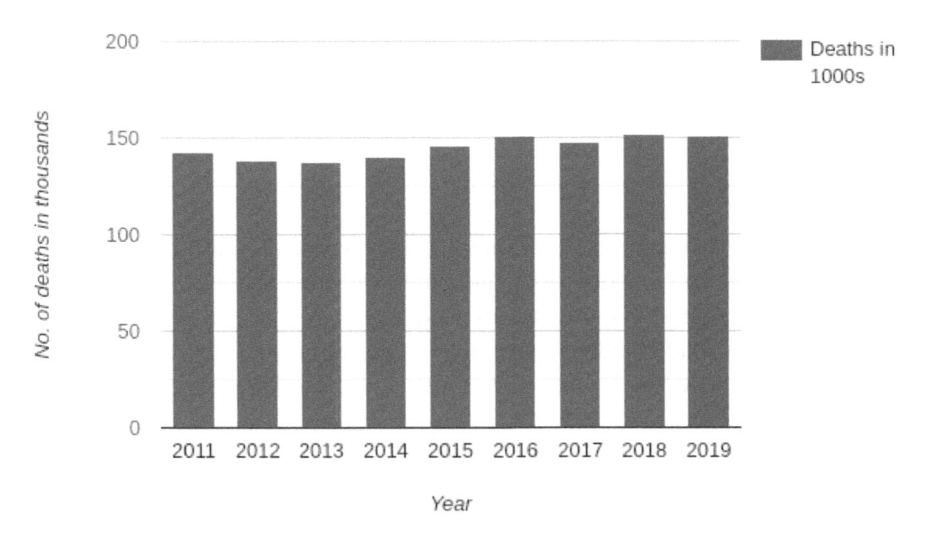

Figure 4.2 The number of deaths in thousands.

The proposed system will continuously monitor the driver's visual information like facial landmarks and neck movements such as rotation and flexion to detect drowsiness and distraction using artificial intelligence [4]. With the significant advances in miniaturization of required electronic equipment and the usage of microelectronics in vehicle systems, the prototype hardware system will be equipped with ultrasonic sensors [5] on all its sides to detect close objects connected to an alarm system to alert the driver using a Bluetooth module [6] to validate the proposed model. An automatic braking system will be triggered if the driver cannot initiate brakes, halting the vehicle to stop minimizing the destruction caused.

4.2 Literature Review

The approach in [7] does not include vertical head movement, which can be helpful in determining whether or not a person has drifted off. Furthermore, little emphasis was placed on alerting the driver when drowsiness was detected, which could have been implemented.

In [8], collision on the rear side of the vehicle has not been accounted for, which also holds equal likelihood. Furthermore, [8] only gives features helpful in driving safety, but it does not provide an algorithm for implementing these features.

In [9], infrared sensors are used for detection purposes, which are less accurate than the computer vision algorithm that we employ. The work in [9] does not emphasize driver safety if he falls asleep, which will be discussed further in this chapter.

4.2.1 Motivation

It's worth noting that the values in Table 4.1 are consistent year after year. This fact pushes us to consider what outcomes we have achieved thus far in terms of driver safety. It motivates us to design a robust and promising system or effective road accident prevention system to ensure driver safety and a considerable reduction in accident statistics. Before jumping right into

Table 4.1 Road accidents and deaths caused data.

Year	No. of road accidents	Deaths (in 1000s)
2011	497,686	142.49
2012	490,383	138.26
2013	486,476	137.57
2014	489,400	139.67
2015	501,423	146.13
2016	480,652	150.79
2017	464,910	147.91
2018	467,044	151.42
2019	449,002	151.51

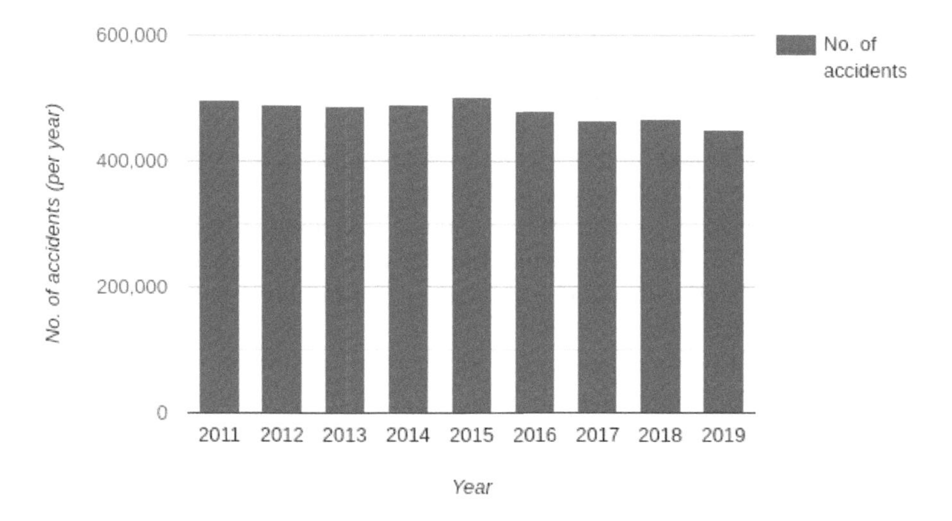

Figure 4.3 The number of deaths in thousands.

the remedy, we must first understand the problem, the reason or source of these incidents, as this will direct us to the most relevant area to work on.

According to several studies, driver tiredness and lack of focus account for around 20% of all crashes and 30% of fatal crashes. Fatigue is a more obvious phrase for this type of circumstance. Fatigue impairs a driver's perception and decision-making abilities to maintain proper control of the vehicle. According to research, a driver's steering performance deteriorates after 2–3 hours of continuous driving [10]. According to recent studies, deploying driver face monitoring devices will lower the number of crashes by 10% to 20%.

Despite the fact that many articles and research papers on safety systems have been published, there is still room for development in either the algorithms or the overall preventative concept.

4.2.2 Contribution

The distraction or drowsiness detection and prevention algorithms play a significant role in a driver safety system in vehicles. Tracking the relative change in position of the driver's facial landmarks [11] between successive frames is highly helpful in predicting facial expressions and conditions more precisely and with a minor error. The dataset used in the model is downloaded from the Intelligent Behaviour Understanding Group (iBUG) repository.

Here are the contributions of the proposed idea and model:

1. The relative positions between various facial landmarks are compared in the model, like landmarks around eyes, mouth, nose, and face boundaries. The algorithm's speed, accuracy, and complexity in each result are analyzed with various performance metrics.
2. Image enhancement techniques like contrast stretching [12], histogram equalization [13], log, and gamma transformations [14] are compared based on their efficiency and error rate in performing face detection algorithms in the image.
3. The novelty of work lies in developing an algorithm to predict drowsiness and distraction using calculation on the relative position of facial landmarks, and further triggering different warnings based upon the results obtained.

Our significant contributions are described as follows:

1. The accurate prediction of distraction and drowsiness using image enhancement and facial landmarks detection techniques is improved.
2. The precision of the results obtained is improved.
3. The time complexity and latency optimization in the algorithm.

4.2.3 Organization

In this chapter, Section 4.3 describes the detailed description of the methodology and workflow of our proposed model. Section 4.4 defines the performance of the proposed model along with evaluation parameters. Finally, Section 4.5 discusses the conclusion of this work.

4.3 Proposed Methodology

We can quickly identify faces and predict people's facial exprcssions in photos or videos, but making any code or algorithm do that for you will be extremely useful. Using this information, we can create powerful applications. However, we need first to detect the face and retrieve information about the face, like whether the mouth or eyes are opened or closed. For this, various proposed face recognition techniques or methods have been explored [15] [16]. This face detection technology is used to detect faces and facial expressions from digital images and videos [17]. So, using this information, we can predict whether a person smiles or laughs. Also, critical information about the driver's facial expression and movement can help us predict distraction or drowsiness. In this proposed model, using facial landmarks, one can determine the different facial expressions of people out of it. We used one of the most potent algorithms like HOG and linear SVM, which gives you 68 points (landmarks) of the face. The conditions that could be predicted from the driver's facial expressions are *Distraction*, *Yawning*, and *Dozing Off*.

The flow of proposed model:

1. The video is recorded live in the form of images, frame by frame.
2. The image is then converted to grayscale [18] to get the coordinates of faces present in the image.
3. The coordinates are classified using histogram of oriented gradients and linear SVM.
4. Each grayscale video frame is enhanced using contrast stretching.
5. The relative change between these landmarks is tracked using their position between successive frames.
6. The prediction of distraction, yawning, and dozing are analyzed using distances between facial landmarks coordinates.

Figure 4.4 Attentive and awake person image.

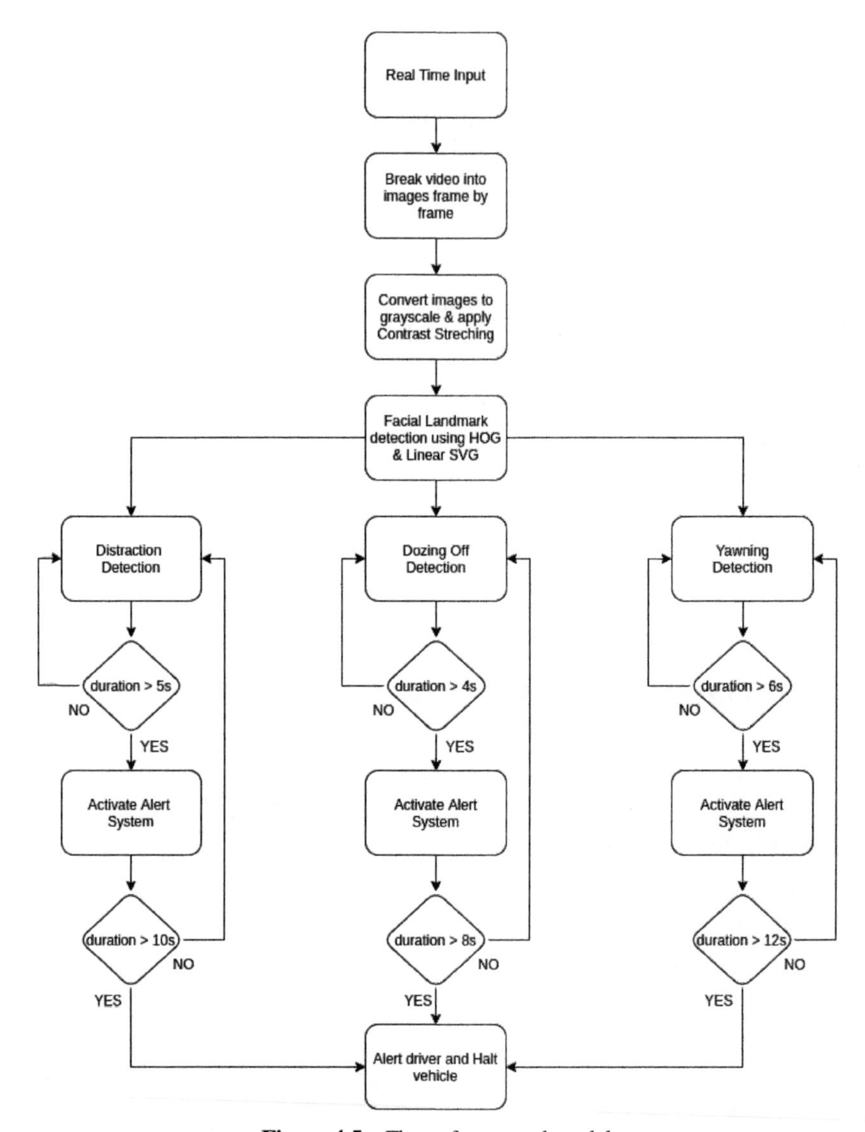

Figure 4.5 Flow of proposed model.

4.3.1 Landmark detection

The video is captured live frame by frame in the form of images. The image is then converted to grayscale so that the pre-trained model combining histogram of oriented gradients (HOG) and linear SVM trained can predict the facial landmarks available from the frames.

HOG is a feature descriptor algorithm frequently used to extract features from image data by extracting the gradient and orientations calculated in "localized" portions.

Linear SVM is a popular algorithm that creates a line or a hyperplane [19] and separates the data into classes, and further finds the points closest to the line from both classes. These points obtained are called support vectors. Now, SVM computes the distance between the line and the support vectors. This computed distance is called the margin, which is to be maximized. Thus, SVM tries to make a boundary so that the separation between the two classes (that street) is as wide as possible.

4.3.2 Contrast stretching

It is a simple image enhancement technology that improves the contrast in an image by expanding the range of intensity values to the full range of pixel values shown in Figures 4.6–4.9. This is one of the techniques most frequently used to enhance nighttime images or videos [20]. We worked on 8-bit pixels in video frames in this model; we could perform contrast stretching between 0–255 range of values. Each input pixel contrast (Xinput) which lies between [Xmin, Xmax] range, is mapped to [0, 255] as follows

$$X_{\text{new}} = 255 * \frac{X_{\text{input}} - X_{\text{min}}}{X_{\text{max}} - X_{\text{min}}}. \tag{4.1}$$

4.3.3 Data storage

The separate time attributes are created for three facial conditions of distraction and drowsiness, respectively. The time attributes used are:

1. Start time, stores the time of the first detection of facial condition either since the start of the driver's face video capturing or after appreciable time since time reset of the same condition.
2. Current time, stores the current time of facial condition.
3. Time elapsed, stores the time difference between the current time and start time.

4.3.4 Distraction detection

The designed model will detect distraction only for the vehicles in motion because distraction detection is redundant. The estimation of the vehicle's motion could be estimated from the acceleration using an accelerometer or speedometer [21]. The proposed model assumes that the alignment of the

Figure 4.6 Original RGB image.

Figure 4.7 Enhanced RGB image.

Figure 4.8 Original black and white image.

Figure 4.9 Enhanced black and white image.

face and camera would be the same for accurate predictions. The proposed model checks for any presence of bounding box around the face. If the person is not looking straight into the camera, can initiate distraction detection. Also in the worst case if video frame captured is not of good quality or one of the components of the system is faulty can make detection difficult. So to avoid these issues the time period component is important to properly classify a case of distraction. Upon detecting distraction, the level-wise alert will be triggered based on the duration of distraction. The steps evolved are:

1. Distraction time elapsed >5 seconds: Level 1 warning is activated, which involves triggering an alarm system to alert the driver and ultrasonic sensors for obstacle detection. If an obstacle is found, the braking sequence is activated, and an alarm will be turned off once the vehicle has come to rest.
2. Distraction time elapsed >10 seconds: Level 2 warning is activated, which also involves triggering alarm system to alert the driver, and braking sequence will be applied irrespective for obstacle detection, and an alarm will be turned off once the vehicle has come to rest.

4.3.5 Yawning detection

The yawning detection is calculated by assuming a line between landmarks 61 and 65, find the perpendicular distance between upper and lower lips i.e, between landmarks 63 and 67. We also find distances between landmarks 61 and 65 as shown in Figure 4.12. Finally,

$$factor_1 = \frac{d_1 - d_2}{d_3}, \tag{4.2}$$

where

d1 = Perpendicular distance from 63 to line joining 61 and 65
d2 = Perpendicular distance from 67 to line joining 61 and 65
d3 = Distance between landmarks 61 and 65
For yawing detection, the $factor_1$ value should be greater than 0.15.

1. Yawning time elapsed >6 seconds: Level 1 warning is activated, which triggers an alarm system to alert the driver.
2. Yawning time elapsed >12 seconds: Level 2 warning is activated, which involves triggering alarm system to alert the driver and ultrasonic sensors for obstacle detection. If an obstacle is found, the braking sequence is activated, and an alarm will be turned off once the vehicle has come to rest.

Figure 4.10 Distracted person image.

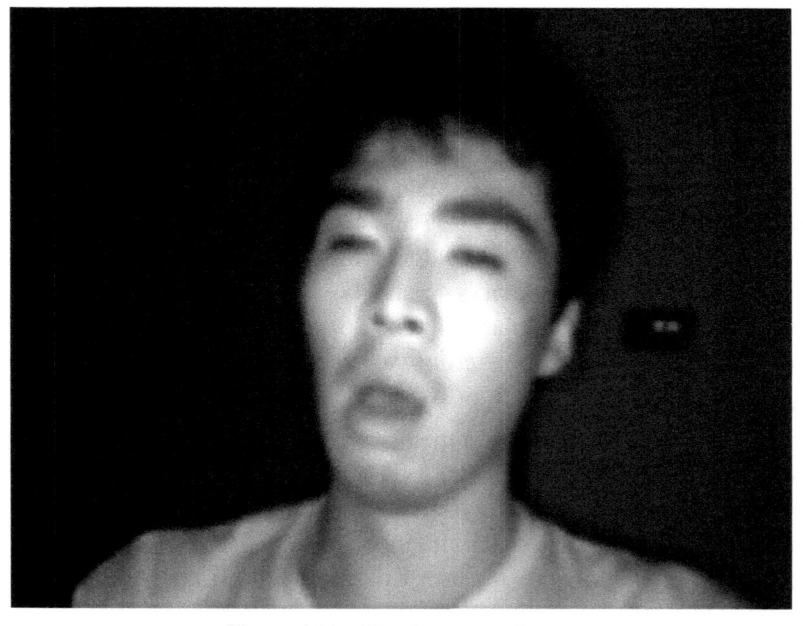

Figure 4.11 Yawning person image.

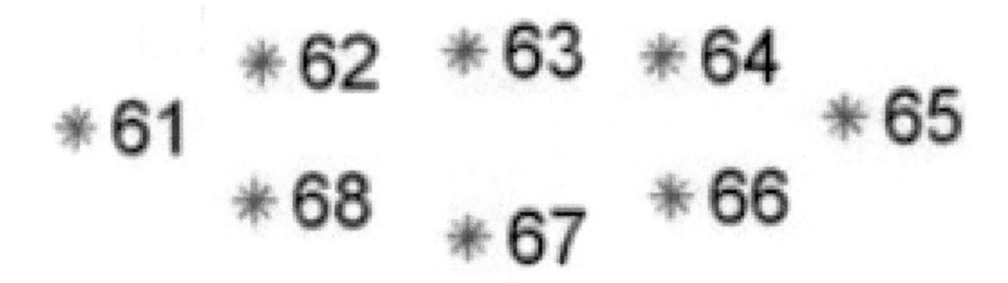

Figure 4.12 Landmarks around mouth region.

4.3.6 Dozing-off detection

The states of both eyes are used for predicting dozing-off detection. First, for the left eye, construct a line between landmarks 37 and 40. Then, find the distance between eyelids, i.e., landmarks 38 and 42. Similarly, find the distance between landmarks 39 and 41. Divide the last two distances by the distance between landmarks 37 and 40 to normalize the value across frames. Similarly, performing the same operations for the right eye. Finally,

$$factor_2 = \frac{\frac{d_4 - d_5}{d_6} + \frac{d_7 - d_8}{d_9}}{2}, \tag{4.3}$$

where

d4 = Distance between landmarks 38 and 42
d5 = Distance between landmarks 39 and 41
d6 = Distance between landmarks 37 and 40
d7 = Distance between landmarks 44 and 48
d8 = Distance between landmarks 45 and 47
d9 = Distance between landmarks 43 and 46

For dozing-off detection, the $factor_2$ value should be smaller than 0.27.

1. Dozing time elapsed >4 seconds: Level 1 warning is activated, which triggers an alarm system to alert the driver.
2. Dozing time elapsed >8 seconds: Level 2 warning is activated, which involves triggering an alarm system to alert the driver and ultrasonic sensors for obstacle detection. If an obstacle is found, the braking sequence is activated, and an alarm will be turned off once the vehicle has come to rest.

Figure 4.13 Dozing-off person image.

*44 *45 *38 *39
*43 *48 *47 *46 *37 *42 *41 *40

Figure 4.14 (a) Landmarks around left eye region (b) Landmarks around right eye region.

4.4 Experimental Results

To analyze the performance of the proposed model, various evaluation parameters are defined such as confusion matrix, accuracy, precision, recall and specificity and F1 score. These parameters are being briefly discussed in the upcoming subsections.

There are a few common terminologies to be aware of, which are:

TP = number of true positive decisions

TN = number of true negative decisions

FN = number of false negative decisions

FP = number of false positive decisions

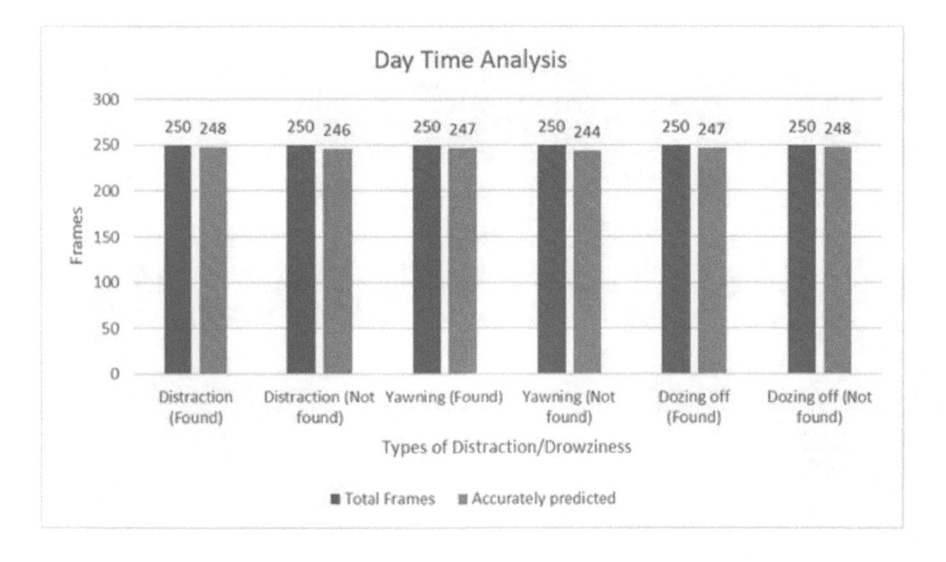

Figure 4.15 Comparison of results during day time.

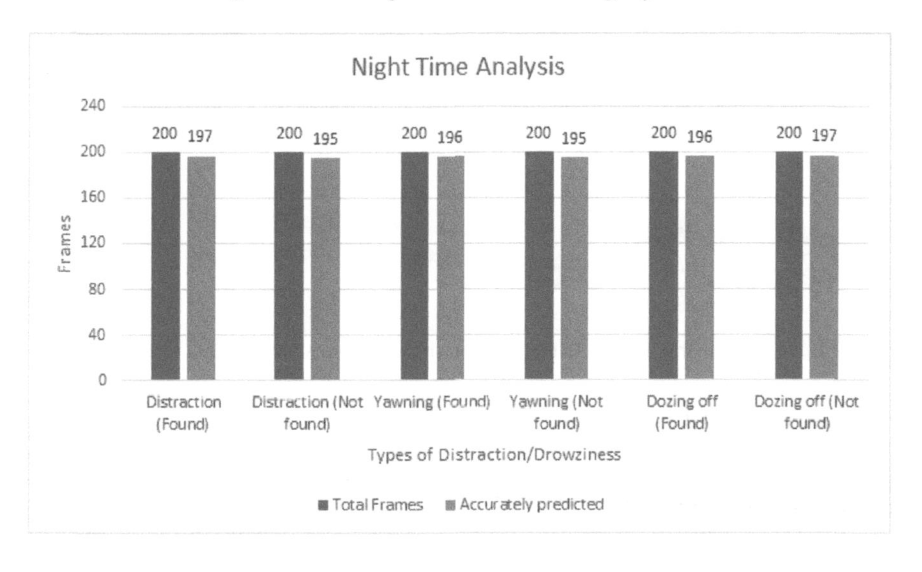

Figure 4.16 Comparison of results during day time.

Table 4.2 Confusion matrix for day time analysis.

	Positive	**Negative**
Positive	742	8
Negative	12	738

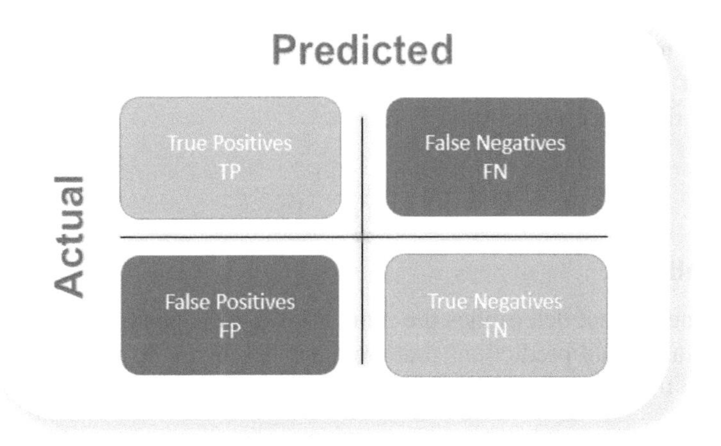

Figure 4.17 Confusion matrix.

Table 4.3 Confusion matrix for night time analysis.

	Positive	**Negative**
Positive	589	11
Negative	13	587

Table 4.4 Performance analysis of proposed model.

Type	**Accuracy**	**Precision**	**Recall**	**Specificity**	**F1 score**
Day	0.987	0.984	0.989	0.984	0.986
Night	0.980	0.978	0.982	0.978	0.980

4.4.1 Confusion matrix

The findings of the prediction are summarized in a confusion matrix [22]. The number of accurate and wrong predictions is totaled and split down by class using count values. The confusion matrix's key is this. The classification model is confused when it makes predictions, as seen by the confusion matrix.

4.4.2 Accuracy

Accuracy is a regularly used metric to determine the correctness of the model predictions, however it is not a reliable indicator of performance. When classes are unbalanced, the situation becomes even worse. Accuracy is calculated as follows

$$\text{Accuracy} = \frac{\text{TP} + \text{TN}}{\text{TP} + \text{FP} + \text{FN} + \text{TN}}. \tag{4.4}$$

4.4.3 Precision

Precision is a metric that quantifies the number of correct positive predictions made out of all positive predictions. Therefore, precision calculates the accuracy for the minority class.

$$\text{Precision} = \frac{\text{TP}}{\text{TP} + \text{FP}}. \tag{4.5}$$

4.4.4 Recall

Recall is a metric that determines the number of correct positive predictions made out of all actual predictions that are made. In this way, recall provides some idea on the coverage of the positive class.

$$\text{Recall} = \frac{\text{TP}}{\text{TP} + \text{FN}}. \tag{4.6}$$

4.4.5 Specificity

Specificity is a metric that estimates the number of correct negative predictions made out of all actual negative predictions that are made. In this way, specificity provides some idea on the coverage of the negative class.

$$\text{Specificity} = \frac{\text{TN}}{\text{TN} + \text{FP}}. \tag{4.7}$$

4.4.6 F1 Score

F1 score is the harmonic mean of precision and recall. Because this metric takes into account both, the greater the F1 score, the better. As you can see, lowering any of the factor in the numerator lowers the ultimate F1 score significantly. If the positive predicted are actually positives (precision) and doesn't miss out on positives and predicts them negative (recall), a model does well in F1 score.

$$\text{F1}_{\text{score}} = \frac{2}{\frac{1}{\text{precision}} + \frac{1}{\text{recall}}} \tag{4.8}$$

or

$$\text{F1}_{\text{score}} = \frac{2 \cdot \text{precision} \cdot \text{recall}}{\text{precision} + \text{recall}}. \tag{4.9}$$

4.4.7 Comparison with some state-of-the-art methods

1. PERCLOS and Linear SVM [23]: This technique used one of the key drowsiness and fatigue detection methods called PERCLOS (percentage of eye closure) and eye aspect ratio (EAR). PERCLOS involves the calculation of the ratio between the number of frames in which drowsiness is detected divided by the number of total frames. It used linear SVM to predict facial landmarks around the eyes, similar to the proposed method. But the number of frames method involved in PERCLOS leads to the dependency of the algorithm to computational speed. So, less powerful devices can take a long time to predict drowsiness properly. Even if the driver is drowsy, it will cost a valuable time in which he is actually alerted. Also, even though eyes are critical in any drowsiness algorithm, it failed to consider the detection of a yawn, which is helpful in alerting the driver of drowsiness so that the driver can take the necessary action to overcome the situation.
2. Kernelized Correlation Filter and Multiple CNN [24]: This drowsiness detection algorithm uses the coordinates of facial landmarks around the eyes and mouth and further creates a bounding box over them. This algorithm used CNN to predict drowsiness rather than relying on simple yet effective mathematical computations since CNN requires a dataset to be trained. But it provided an accuracy of 95%, which is close to the perfect model.
3. Multilayer Perceptron Classifier [25]: This algorithm performs drowsiness prediction based on the position of all 68 facial landmarks. The values of coordinates are trained using multilayer perceptron classifier. The use of facial landmarks except around eyes and mouth can cause loss of a valuable amount of accuracy and may result in overfitting over the training dataset. Also, as there are no thresholds set on how the model should behave in corner situations, the drowsiness detection model can trigger false warnings to the driver.
4. Multilayer CNN using AlexNet [26]: This algorithm used multilayer CNN to detect drowsiness directly on frames of video samples without detecting any facial landmarks on the image. This can cause high overfitting over the training dataset unless a very large amount of drowsiness dataset is available, and even then, it will not guarantee a good accuracy. So, similar to the problem in the algorithm proposed in multilayer perceptron classifier, this model can also trigger false warnings to the driver.

Table 4.5 Comparing some state-of-the-art methods.

Index No.	Algorithms	Average accuracy
1.	PERCLOS and linear SVM	0.948
2.	Kernelized correlation Filter and multiple CNN	0.950
3.	Multilayer perceptron classifier	0.809
4.	Multilayer CNN using AlexNet	0.808
5.	CNN using ResNet	0.880
6.	Ours	0.984

5. CNN using ResNet [27]: This algorithm predicts distraction based upon deep neural networks. Multiple cameras are used where front cameras are used to record face movements and side cameras to record hand and body movements. These recorded videos are trained using deep neural networks. Although head movements are crucial for detecting distraction but recording by the side cameras can falsely classify some hand and body movements as distractions. So instead of movements driver steering or changing gears, it can classify these actions as "drinking" or "texting on the phone". Also, the classification is being done frame by frame but not based on some actions over a time period which negatively affects the system's accuracy and can trigger repetitive false warnings.

4.5 Conclusion

Safety is the everlasting theme of automobile technology development. With the development of society and economy, car ownership keeps increasing; urbanization promotes the change of urban and rural traffic environment, and the problem of automobile safety becomes more and more serious. This paper proposes an improved model depicting higher face detection accuracy in day and low light conditions. The automobile prototype for physical simulation will be equipped with ultrasonic sensors communicating with the automatic halting system using a Bluetooth module and an alerting system to notify the driver and store user activity data. Then, further improvements can be made according to the analysis results.

References

[1] Vikram Shende. Analysis of research in consumer behavior of automobile passenger car customer. *Iinternational Journal for Scientific and Reaseach Publications*, 4, 02 2014.

[2] George Yannis, Eleonora Papadimitriou, and Katerina Folla. Effect of gdp changes on road traffic fatalities. *Safety Science*, 63:42–49, 03 2014.

[3] Ramesh Kasimani, Muthusamy P, Rajendran M, and Palanisamy Sivaprakash. "a review on road traffic accident and related factors". *International Journal of Applied Engineering Research*, 10:28177–28183, 01 2015.

[4] Charlotte Jacobé de Naurois, Christophe Bourdin, Anca Stratulat, Emmanuelle Diaz, and Jean-Louis Vercher. Detection and prediction of driver drowsiness using artificial neural network models. *Accident; analysis and prevention*, 126, 12 2017.

[5] Alessio Carullo and Marco Parvis. An ultrasonic sensor for distance measurement in automotive applications. *Sensors Journal, IEEE*, 1:143 – 147, 09 2001.

[6] Rajesh Shrestha. Study and control of bluetooth module hc-05 using arduino uno. 01 2016.

[7] Wanghua Deng and Ruoxue Wu. Real-time driver-drowsiness detection system using facial features. *IEEE Access*, PP:1–1, 08 2019.

[8] Dr D Yuvaraj. Enhancing vehicle safety with drowsiness detection andcollision avoidance. 04 2017.

[9] Stephen Eduku, Mohammed Alhassan, and Joseph Sekyi-Ansah. Design of vehicle accident prevention system using wireless technology. 10 2017.

[10] Al-Baraa Al-Mekhlafi, Ahmad Isha, Nizam Isha, and Gehad Naji. The relationship between fatigue and driving performance: A review and directions for future research. *Journal of critical reviews*, 7:2020, 07 2020.

[11] G. Beumer, Qian Tao, Asker Bazen, and Raymond Veldhuis. A landmark paper in face recognition. pages 73–78, 01 2006.

[12] Kartika Firdausy, Tole Sutikno, and Eko Prasetyo. Image enhancement using contrast stretching on rgb and ihs digital image. *TELKOMNIKA (Telecommunication Computing Electronics and Control)*, 5:45, 04 2007.

[13] Omprakash Patel, Yogendra Maravi, and Sanjeev Sharma. A comparative study of histogram equalization based image enhancement

techniques for brightness preservation and contrast enhancement. *Signal & Image Processing : An International Journal*, 4, 11 2013.
[14] Guan Xu, Jian Su, Hongda Pan, Zhiguo Zhang, and Haibin Gong. An image enhancement method based on gamma correction. volume 1, pages 60–63, 01 2009.
[15] Alamgir Sardar, Saiyed Umer, Chiara Pero, and Michele Nappi. A novel cancelable facehashing technique based on non-invertible transformation with encryption and decryption template. *IEEE Access*, 8:105263–105277, 2020.
[16] Saiyed Umer, Bibhas Chandra Dhara, and Bhabatosh Chanda. Biometric recognition system for challenging faces. In *2015 Fifth National Conference on Computer Vision, Pattern Recognition, Image Processing and Graphics (NCVPRIPG)*, pages 1–4. IEEE, 2015.
[17] Suja Palaniswamy and Shikha Tripathi. Emotion recognition from facial expressions using images with pose, illumination and age variation for human-computer/robot interaction. *Journal of ICT Research and Applications*, 12:14, 04 2018.
[18] Saravanan Chandran. Color image to grayscale image conversion. pages 196 – 199, 04 2010.
[19] Panos Pardalos. Hyperplane arrangements in optimization. 01 2008.
[20] Zhenghao Shi, Mei Zhu, Bin Guo, Minghua Zhao, and Changqing Zhang. Nighttime low illumination image enhancement with single image using bright/dark channel prior. *EURASIP Journal on Image and Video Processing*, 2018, 02 2018.
[21] Quan Yuan, Tang Shenjun, Liu Kaifeng, and Li Yibing. Investigation and analysis of drivers' speedometer observation and vehicle-speed cognition. pages 667–670, 01 2013.
[22] Kai Ting. *Confusion Matrix*, pages 260–260. 01 2017.
[23] Feng You, Xiaolong Li, Yunbo Gong, Hailwei Wang, and Hongyi Li. A real-time driving drowsiness detection algorithm with individual differences consideration. *IEEE Access*, 7:179396–179408, 2019.
[24] Wanghua Deng and Ruoxue Wu. Real-time driver-drowsiness detection system using facial features. *IEEE Access*, 7:118727–118738, 2019.
[25] Rateb Jabbar, Khalifa Al-Khalifa, Mohamed Kharbeche, Wael Alhajyaseen, Mohsen Jafari, and Shan Jiang. Real-time driver drowsiness detection for android application using deep neural networks techniques. *Procedia computer science*, 130:400–407, 2018.

[26] Sanghyuk Park, Fei Pan, Sunghun Kang, and Chang D Yoo. Driver drowsiness detection system based on feature representation learning using various deep networks. pages 154–164, 2016.

[27] Duy Tran, Ha Manh Do, Weihua Sheng, He Bai, and Girish Chowdhary. Real-time detection of distracted driving based on deep learning. *IET Intelligent Transport Systems*, 12(10):1210–1219, 2018.

5

Texture Feature Descriptors for Analyzing Facial Patterns in Facial Expression Recognition System

Sanoar Hossain[1] and Vijayan Asari[2]

[1]Department of Computer Science & Engineering, Aliah University, India
[2]Electrical and Computer Engineering, University of Dayton, USA
E-mail: snr.hossain12@gmail.com; vasari1@udayton.edu

Abstract

A multipose facial expression recognition system has been proposed in this chapter. The proposed system has three components such as (i) image preprocessing, (ii) feature extraction, and (iii) classification. During image preprocessing, from the input body silhouette image, the facial landmarks detection has been performed. Then using these landmark points the rectangular-box face region is detected. Here both frontal and profile face images have been considered for recognizing the type of facial expression in the image. Then a multi-level texture feature descriptors analysis approach is being employed to analyze the texture patterns in the detected facial region for feature computation. For this multi-level feature computation, various global and local feature representation schemes have been employed to extract more distinctive and discriminant features. Finally, a multi-class classification technique using support vector machine classifier is employed to perform classification task to detect the type of facial expression in the image. The performance of the proposed system has been tested on one challenging benchmark facial expression databases such as Karolinska directed emotional faces (KDEF) database and then the performances have been compared with the existing state-of-the-art methods to show the superiority of the proposed system.

Keywords: Bag-of-words model, facial expression, multi-level, multi-pose, recognition, support vector machine, texture analysis.

5.1 Introduction

Human facial expression recognition (FER) and emotions classification is one of the hot topics and most emergent research area in image processing, pattern recognition, machine learning, and computer vision to understand human emotions [1] because of its large potential applications and theoretical challenges. Face recognition has great attention in biometric, cosmetic product, tourist satisfaction, health management, computer game design, and in image analysis and synthesis application. Facial expression has an important role in our daily communication with people [2] and social interaction [2]. It is an explicit indication of the affective state, intellectual movement, personality, self-aided driving, attention and indication of mind. For centuries, researchers have studied on face expression and emotion in the field of social physiology and physiology [3] and exhibited that speech signal and facial expression are more informative for human emotion recognition. Albert Mehrabian *et al.* [4] had proposed that 55% emotional information among total information has been transmitted through facial expression in daily human interaction while voice contributes 38%, and language accounts seven percent. Though it is an easy task for human beings to recognize facial expression, it is quite tough to be performed by a computer. It is somewhat quite difficult to boost a robotized framework that can decipher emotion and recognized facial expression. It is an immediate powerful and effective nonverbal-way of communication to transit message and convey emotional information [2]. FER is considered to be a significant part of natural user interface. Human brain can easily identify emotional expressions by looking at the facial muscular movement, contracted and characteristics. Here, the psychology of emotion [5] has considered it as the unique important gesture. Lately, facial expression recognition system (FERS) and emotion classification are used in the diversified field of application like e-Healthcare and m-Health [34] to understand the patients' current state even when the patients are too far from doctors, nurses, and hospitals. FERS is also used to understand the patient's pain whether he / she is feeling high, low or moderate. or no pain when he/ she can't talk [35].

Automatic facial expression recognition and classification of multi-pose and multi-level face images have turned out to be an attractive and most challenging problem since the last 30 years [8]. A literature review states that early stages of research had focused on several statistical and structural based method for facial expression recognition [8]. Here, both template-based and feature-based methods were observed. Recently, researchers have developed many automatic facial expression recognition and classification system with the help of deep learning framework, video data, and image sequence. Aiming toward the applications of people sentimental analysis and human-machine or computer interaction, this topic has recently drawn even more attention. Facial expressions are mainly used for face recognition task [2] which is an image classification problem [2]. Basically, the face image $\mathcal{I}$ contains the seven basic emotion classes or expressions e.g., fear (AF), anger (AN), disgust (DI), happy (HA), neutral (NE), sad (SA), and surprise (SU) and these expressions are considered for experimentation. Figure 5.1 shows some examples of emotion expressions.

Facial expression recognition has two main approaches. They have been analyzed: appearance-based methods and geometric-based methods [9]. Xijian Fan *et al.* [10] used geometric-based appearance feature and texture features to propose an automated deep learning facial expression recognition framework. The appearance-based methods are the most successful and well-studied for face recognition. In this work, the whole face image captured

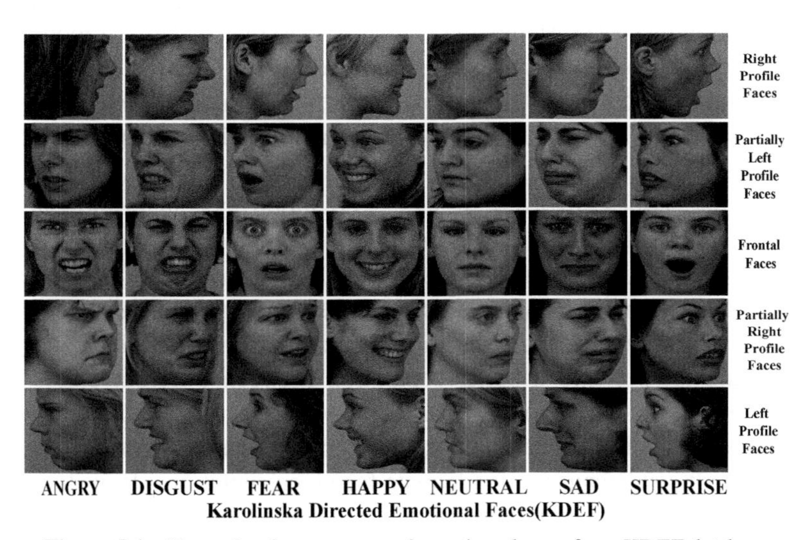

Figure 5.1 Example of seven types of emotion classes from KDEF database.

in under controlled-lab condition is taken as the input image $\mathcal{I}$ to create a subspace based on the reduction of inconsistent and redundant face space dimensionality using various subspace methods [11] such as Fisher LDA, PCA, and LPP [17] [13]. A comparative literature review of these techniques for facial emotion expression recognition and classification could be seen in [14] and [15]. In subspace-based method, face manifold may or may not be a linear space [16].

Since LDA and PCA practically see only the kernel methods [17], Euclidean structure and miscellaneous learning methods [18] have been employed for face recognition. The computation cost of these techniques is expensive and some times this system may be a failure to explicitly exhibit the exact structure of the manifold. But this is a powerful tool for statistical signal modeling that is known as sparse coding. Sparse coding provides beautiful results for facial expression recognition [19]. Since in our proposed methodology, we have considered that the facial expression recognition is an image classification and pattern recognition problem, we have considered that the face image of the same person would belong to the same class or category. The feature descriptors are extracted from each facial image and built a pyramid of histogram to represent each category of image. LBP [20], HOG [60] [22] [23], and spatial dense SIFT [23] [24] features method followed by BoW(SIFT) have been employed to recognized categories and face representations. Instead of using non-linear support vector machine (SVM), we have employed linear support vector machine for the final classification task [25]. Hence the choice of classifier makes the model more robust and efficient in kernel space. Our extensive experimental results in Table 5.3 and Table 5.4 demonstrate that the proposed method can achieve excellent recognition accuracy for multi-pose and multi-level images on KDEF databases.

The principle issues involved in face recognition system design are face representation and classifier selection [17]. Face representation involves extraction of feature descriptors from the input face image and it would minimize intra-class similarities and maximize the inter-class dissimilarities. In case of classifier selection, it doesn't make sense that the high-performance classifier would always find a better separation between different classes even if there are some significant similarities between each other. Sometimes the most sophisticated classifier fails to execute the face recognition and classification task just because of inadequate face representations. If we have employed good face representation but don't select a proper classifier, we can't achieve high performance recognition accuracy. Recently, basic facial expression detection and emotion recognition problem were solved under controlled lab conditions for occlusions, posed variation, and frontal facial expression accomplishing better accuracy. But the problem of FER is still challenging for different profile faces, inter-and intraclass pose variation under illumination and low resolution conditions. The FERS problems have been solved using advanced data augmentation methods with deep feature learning technique [35] and advance hyper-parameters tuning, transfer learning, score-level fusion techniques, and ultra-deep convolution neural network (CNN) [37].

Numerous state-of-the-art for multi-level and multi-pose facial expression recognition and classification systems used hand-labeled points and takes advantages for feature extraction [28]. Rao *et al.* [29] proposed a novel local feature extraction technique from diverse overlapping patches boosting speeded-up robust feature (SURF) extraction algorithm and achieved favorable result on RaFD databases. Xijian Fan *et al.* [10] has proposed an automated deep learning framework to enhance the result of recognition accuracy for facial expression recognition. Here the proposed method combines the discriminative texture features and handcraft features. Convolution neural network (CNN) have been employed for feature learning and feature selection. Linear SVM classifier is used for final image classification. The feature level fusion is used to optimize the recognition rate. The CK++ data is used for experimentation and gives better recognition. Gutta *et al.* [30] proposed a model with an ensemble radial basis function, grayscale image and inductive decision trees and had used FARET database that have four classes (i.e., Asian, Caucasian, African, and Oriental) for ethnicity. Zhang and Wang [31] proposed a method for two-class (White & Asia) race classification. The proposed method used multi-scale LBP texture features and combines 2D and 3D texture feature in multi-ratio. Hence the proposed method had been tested on FRGC v.2 database and reported a 0.42% error rate.

Zhang *et al.* [32] proposed Gabor wavelets-based texture features and geometry-based texture features for facial expression recognition. Bartlett *et al.* [33] proposed a feature level fusion technique and developed an automatic facial expression recognition framework. Thai Son Ly *et al.* [34] have recently proposed a novel deep learning technique for wild facial expression recognition by jointly fusing 2D and 3D facial features. In [35], Rose proposed a facial expression recognition system with low resolution face images and employed log-Gabor and Gabor filters for feature extraction. Wu *et al.* [36] proposed a dynamic facial expression recognition system as a "six class classification problem." Here the proposed method have employed genetic algorithm, Gabor motion energy (GME) filters, and SVM and used a sequence of video images [37]. Gu *et al.* [38] used Gabor feature to describe a facial expression recognition system.

Wu *et al.* [39] had proposed an efficient and robust method on discrete beamlet transforms regularization for effective blind restoration (DBTR-BR) model to solve different categories of infrared facial expression recognition problem. Almaev *et al.* [40] proposed a feature level fusion technique by combining Gabor filters and LBP-TOP [41] and defined a new dynamic Local Gabor Binary Patterns from Three Orthogonal Planes (LGBP-TOP) descriptor. Xusheng Wang *et al.* [42] has proposed a bi-modal feature-based fusion algorithm in which facial expression and speech signal are fused to get optimal performance in human facial emotion recognition. The hybrid feature extraction network model takes video stream i.e., a number of sequential images as input and produces fully connected layer of convolution neural network for the next cyclic neural network. Zhi *et al.* [43] used a dynamic temporal 3DLeNets model for video analysis to proposed an action unit (AU) enhanced facial expression recognition system [44].

The organization of the proposed work is as follows: The proposed facial expression recognition system (FERS) have been discussed in Section 5.2 which describes multi-level and multi-texture feature extraction techniques for both frontal and profile face region and the classifier design for the classification techniques. The experimental results and discussion has been described in Section 5.4. Finally, Section 5.5 concludes this paper.

5.2 Proposal Methods

Looking at the importance of face recognition in the field of image processing-pattern recognition and computer vision, we propose a robust, efficient, and accurate model for face verification and identification.The proposed facial expression recognition system (FERS) has three components: face region detection, feature extraction, facial expression recognition. The block diagram of the proposed system has been demonstrated in Figure 5.2.

5.2.1 Preprocessing

The input face images are normalized to coequal in size, and eye locations and tip of nose in each image are mapped to the same locations. Preprocessing is a general phenomena that performs for operations at the lowest level

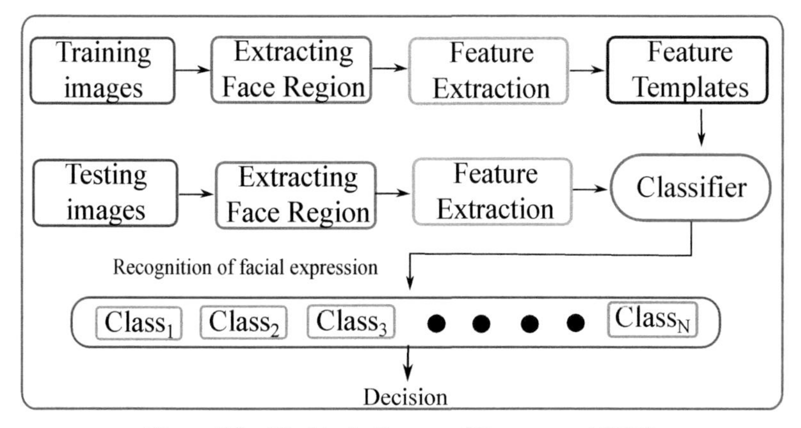

Figure 5.2 The block diagram of the proposed FERS.

of abstraction where both input and output are intensity images. These iconic images are of the same kind as the original data captured by the sensor, with an intensity image usually represented by a matrix of image function values (brightnesses). The aim of preprocessing is an improvement of the image data that suppresses unwilling distortions or enhances some image features important for further processing, although geometric transformations of images (e.g., scaling, rotation, translation) are classified among preprocessing methods here since similar techniques are used [14]. Image preprocessing methods are classified into four categories according to the size of the pixel neighborhood that is used for the calculation of a new pixel brightness: pixel brightness transformations, geometric transformations, certain preprocessing methods use a local neighborhood of the processed pixel and image restoration that requires knowledge about the entire image. Image preprocessing methods use the considerable redundancy in images. Neighboring pixels in local feature extraction corresponding to one object in real images have essentially the same or similar brightness value, so if a distorted pixel can be picked out from the image, it can usually be restored as an average value of neighboring pixels [14].

During preprocessing we extract face region from each input image $\mathcal{I}$ both in global and local feature extraction techniques. Since the facial expressions contain very minute details, so, it is important to be conscious about the analyzing for both expressive or non-expressive characteristics of facial region. During face preprocessing we have applied a tree structured part model [32] which works better for both frontal and profile face region. It has outstanding performance than the other face detection algorithm in computer vision. The tree structured part model works on the principal of mixture of trees with global mixture of topological view points changing. For mixture of trees the face detection algorithm uses templates and these templates are HOG descriptors at the particular pixel location on face region. Hence tree structure part model computes thirty nine (39) land marks point for profile faces while sixty eight (68) landmark points for frontal face. These landmarks points are used to compute four corner points of the face region $\mathcal{I}$. The face preprocessing steps are shown in Figure 5.3.

5.2.2 Feature extraction

Feature extraction is a task to extract discriminating features from the image $\mathcal{I}$ such that the extracted feature must contain more distinctive patterns [45]. In the image processing research areas the feature extraction starts from an initial set of measured data and builds the features which are supposed to be informative and non-redundant, facilitates the subsequent learning and

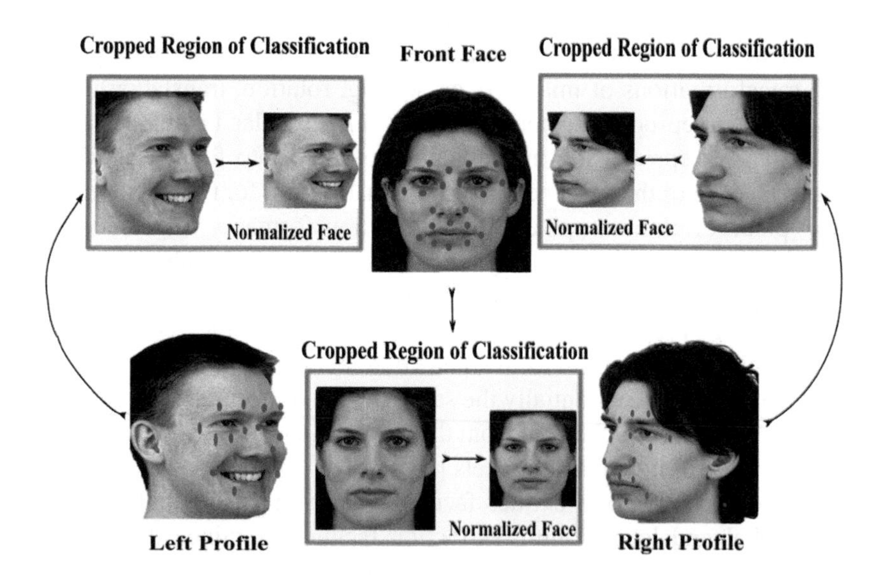

Figure 5.3 The face preprocessing technique for the proposed system.

generalization steps. In many cases the feature extraction techniques lead to better human interpretations. Moreover, it is related to the dimensionality reduction i.e., when the input image is too large to be processed as the representation for that image, then it is transformed into a reduced set of features also called feature vector. There are various feature extraction techniques which are well known in the research area of FER system, and these are as follows:

- *LBP*: Local binary pattern (LBP) [20] is a non parametric feature extraction method that assembles local special texture structure from the image $\mathcal{I}$. It works on the principle of labeling the pixels of images by thresholding a 3×3 neighborhood of each pixel with the center value and considering the results as a binary representation and returns the center element of 3×3 pixels by the gray code of binary representation. Finally histogram of these gray codes over the whole image is the feature vector for $\mathcal{I}$ which is 256 dimensional feature vector.
- *HoG*: Earlier HoG (histogram of oriented gradients) feature is used to retain both shape and texture information [60]. Next it is successfully applied for human detection and now a days it has been emergent applied to face recognition. HoG is similar to scale-invariant feature transformation (SIFT) descriptor [46] [22][23]. HoG feature descriptors are computed block wise [60] using Sobel filters to form horizontal and vertical gradient maps. Magnitude and orientation is computed using horizontal and vertical gradient maps. An image $\mathcal{I}$ of size 128×128 image and feature vector computed from $\mathcal{I}$ is 81.
- *SIFT*: Scale-invariant feature transform (SIFT) [22] algorithm is based on combination of difference-of-Gaussian (DoG) filters, interest region detector and a histogram of gradient (HoG). The descriptors also called key points [24]. Here from each $\mathcal{I}$ of size 128×128, 128 dimensional feature vector is computed.

We use the following methods to extract local and global features from the image $\mathcal{I}$. Texture is an intuitive concept. It is easily observed by people. Texture defies the construction of an appropriate definition [47]. Texture is related to the spatial organization of colors, blur paint and intensities e.g., the blades of grass in a lawn, the spots of a tiger or leopard and the grains of sand in beach. This texture patterns are analyzed in facial images to extract more discriminant and distinctive features. During feature computation, all low-level to higher level of feature abstractions are considered from the images.

Here the first crucial step is to extract low-level visual feature from certain informative region of interest of the input image. Perform some pixel operations on these relative areas of those region of interest based on some visual properties. Features extracted from these relative areas are commonly referred to as local and global features. Hence global features are extracted over the entire image, reflecting some global properties and characteristics of the image by contrast. We do not partition the image $\mathcal{I}$ into blocks or patch. We take the whole image as a single block. We have chosen patch size $n \times n$ similar to the input image $\mathcal{I}_{n \times n}$ size.

Local feature is computed over relatively small region w of the image. Then from each w, the local feature vectors are computed as descriptors. Here we consider $\mathcal{M}$ training samples and then from each sample, l number of the SIFT descriptors are computed densely. Now from some m random number of training samples, we have computed $L = \{l_1, l_2, \cdots l_m\}$ descriptors. Now apply bag-of-words (BoW) [48] model for dictionary learning to obtain $C \in \mathbb{R}^{128 \times \alpha}$, α be the number of cluster centers. Now during feature computation, the SIFT descriptors from each patch $w \in \mathcal{I}$ undergoes to C and obtain the occurrences of descriptors in C in terms of histogram $h \in \mathbb{R}^{1 \times \alpha}$. Finally, features computed from each w, are concatenated together for all β number of patches and obtain the final feature vector $f = (h_1, h_2, \cdots, h_\beta) \in \mathbb{R}^{1 \times \alpha \times \beta}$. These feature extraction techniques have been performed for each block-level partitioning techniques i.e., $\mathcal{P}_0$ (Level-0), $\mathcal{P}_1$ (Level-1), $\mathcal{P}_2$ (Level-2), and $\mathcal{P}_3$ (Level-3).

The feature extraction of the proposed system has been shown in Figure 5.4.

5.3 Classification

For classification purpose a linear support vector machine (SVM) classifier has been selected for multi-class classification purpose. Here the linear SVM solves the multi-class classification problems from large scaled datasets. The objective of linear SVM is to learn λ_a linear functions $\{\mathbb{W}_r^T x \| r \in \mathbb{Y}\}$ from a set of training data $\left\{\left(x_i^k, y_i\right)\right\}_{i \doteq 1}^{\mathcal{M}}$, $k \doteq 1 \cdots \mathcal{N}$ and $y_i \in \mathbb{Y} \doteq [1, 2, \cdots, t, \cdots \lambda_a]$, where $\mathcal{M}$ is the total number of training samples & $\mathcal{N}$ is the total number of local blocks or patches, x_i^k is the ith descriptor of BoW (Dense-SIFT) descriptor for k^{th} local patch. So, for each input descriptor r, the class label is calculated using $y \doteq \max_{r \in \mathbb{Y}} \mathbb{W}_r^T x$.

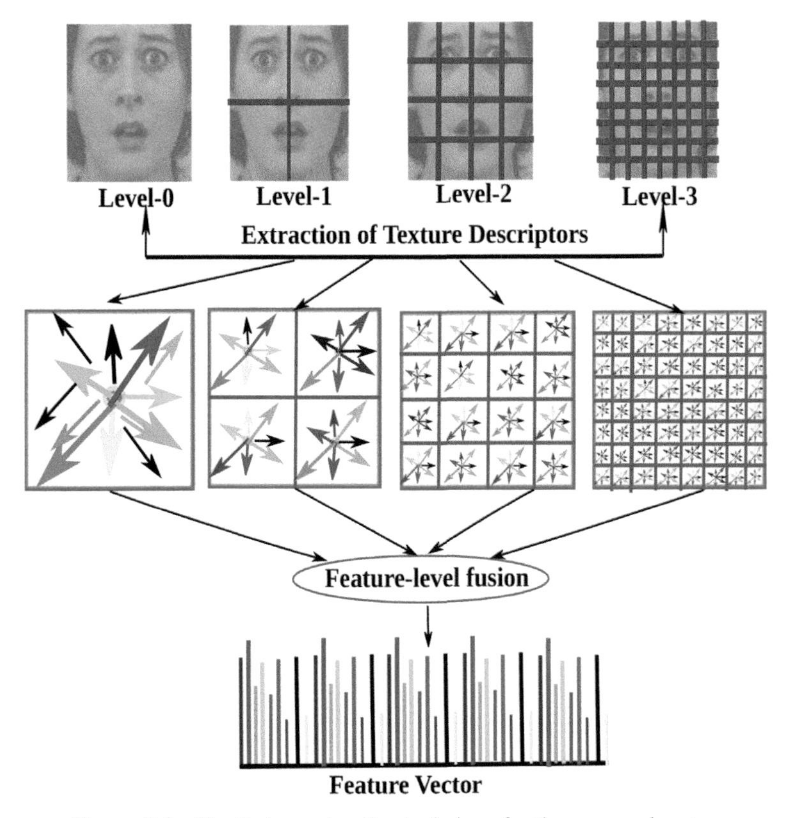

Figure 5.4 The feature extraction technique for the proposed system.

5.4 Experiment and Results

In this section, we have described comprehensive experimental analysis to validate the stability and effectiveness of our proposed method. The proposed method has been experimented on KDEF database and compared its performance with that of some existing state-of-the-art methods.

5.4.1 Database used

For experimental purpose we have selected KDEF (Karolinska directed emotional faces) [1] benchmark databases which contains 70 different individuals (35 male and 35 female) ages between twenty (20) to thirty (30) years. In this work we have used 1210 samples as training-set whereas 1213 samples as testing-set. These images are from seven different facial

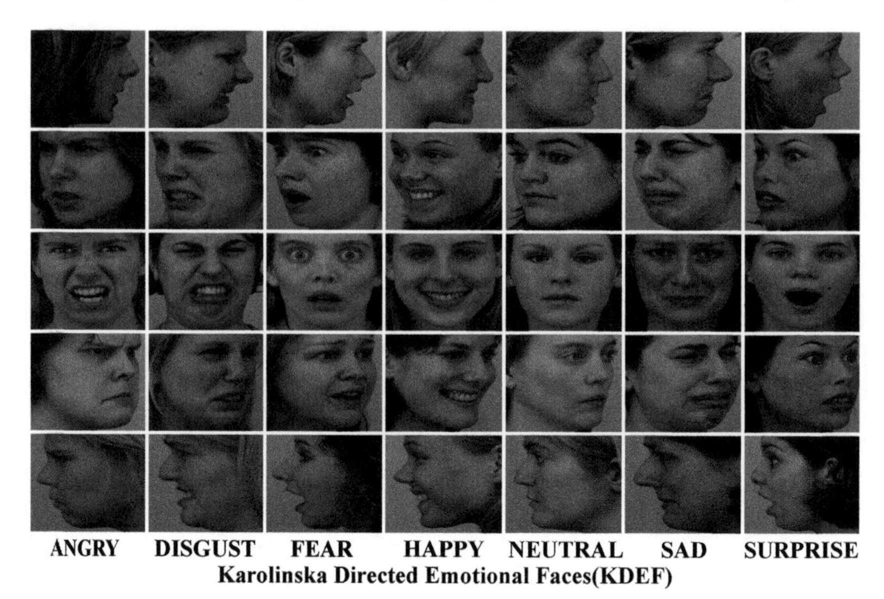

Figure 5.5 Examples from KDEF database.

Table 5.1 Description for facial expression database.

Database	FE Class	Training	Testing
KDEF	7	1210	1213

expression classes i.e., afraid, anger, disgust, happiness, neutral, sadness, and surprise [2]. Moreover, these images contain various challenging issues such as pose variants with five different angles, illumination conditions, ethnicity and geographical locations. Figure 5.5 shows some image samples of the KDEF databases and Table 5.1 summarizes the description of KDEF face databases used for the proposed system.

5.4.2 Result and discussion

The proposed facial expression recognition system (FERS) is implemented using Matlab 16.a version in Windows 10 Pro 64-bit, 3.30 GHz Core-i5 Processor, 8 GB RAM. Here during experimentation we have partitioned $\mathcal{I}_{n\times n}$ into several equal distinct patch $w_{n\times n}$. The input image $\mathcal{I}$ of size 128×128 has been selected. The input image $\mathcal{I}_{n\times n}$ is partitioned into $\mathcal{P}_0, \mathcal{P}_1, \mathcal{P}_2, \mathcal{P}_3$ patches of equal size patches, shown in Table 5.2 at l_0, l_1, l_2, and l_3. Now from each $w_{n\times n}$ from $\mathcal{P}_0, \mathcal{P}_1, \mathcal{P}_2, \mathcal{P}_3$ partition

Table 5.2 Partitioning scheme of $\mathcal{I}_{128\times128}$ for the proposed system.

Partition Scheme	Patch #	Level	Patch size
$\mathcal{P}_0$	2^0= 1	1	128×128
$\mathcal{P}_1$	2^2= 4	2	64×64
$\mathcal{P}_2$	2^4= 16	3	32×32
$\mathcal{P}_3$	2^6= 64	4	16×16

Table 5.3 Performance (in %) of the proposed FERS using various texture features at different image partitioning levels.

Level	$\mathcal{P}_0$		$\mathcal{P}_1$	
Feature	Acc.	Dim.	Acc.	Dim.
LBP	19.26	$\mathbb{R}^{1\times150}$	22.98	$\mathbb{R}^{1\times600}$
HOG	39.50	$\mathbb{R}^{1\times81}$	48.27	$\mathbb{R}^{1\times324}$
SIFT	14.46	$\mathbb{R}^{1\times128}$	14.46	$\mathbb{R}^{1\times512}$
BoW	**39.92**	$\mathbb{R}^{1\times250}$	**50.83**	$\mathbb{R}^{1\times1000}$
Level	$\mathcal{P}_2$		$\mathcal{P}_3$	
Feature	Acc.	Dim.	Acc.	Dim.
LBP	36.12	$\mathbb{R}^{1\times2400}$	45.87	$\mathbb{R}^{1\times9600}$
HOG	55.29	$\mathbb{R}^{1\times1296}$	57.85	$\mathbb{R}^{1\times5184}$
SIFT	32.81	$\mathbb{R}^{1\times2048}$	30.91	$\mathbb{R}^{1\times8192}$
BoW	**59.34**	$\mathbb{R}^{1\times4000}$	**64.79**	$\mathbb{R}^{1\times16000}$

scheme, the features LBP, HOG, SIFT, and BoW (Dense-SIFT) have been extracted accordingly. Here we obtain the perfromance for both 128×128 image size. Finally, these discriminating features from each $w_{n\times n}$ are concatenated to represent as a single feature vector correspond to $\mathcal{I}$. Hence the performance for $f_{\text{LBP}} \in \mathbb{R}^{1\times d_1}$, $f_{\text{HoG}} \in \mathbb{R}^{1\times d_2}$, $f_{\text{SIFT}} \in \mathbb{R}^{1\times d_3}$ and $f_{\text{BoW}} \in \mathbb{R}^{1\times d_4}$ feature vectors are shown in Table 5.3 as accuracy in terms of (%). Here the values for d_1, d_2, d_3, d_4 are shown in Table 5.3.

From Table 5.3 it is observed that for feature vector $f_{\text{BoW}} \in \mathbb{R}^{1\times d4}$, the proposed system has achieved better performance as compare to LBP, HoG, and SIFT features. Further we have shown the performance for both 128× 128 and 256×256 image size by computing feature vector f_{BoW} and these performance has been shown in Table 5.4. From this Table 5.4 it is also observed that the proposed facial expression recognition system has obtained excellent performance for $\mathcal{I}_{128\times128}$.

To achieve better performance, the score-level fusion techniques have been applied on the performance of f_{BoW} for image size 128×128. The fused performance are shown in Table 5.5 and from this Table it is observed that the performance is better for $< \mathcal{P}_2, \mathcal{P}_3 >$ fusion scheme.

Table 5.4 Performance (in %) of the proposed FERS using f_{BoW} features for $\mathcal{I}_{128\times 128}$ and $\mathcal{I}_{256\times 256}$ at different image partitioning levels.

Partition	$\mathcal{I}_{128\times 128}$		$\mathcal{I}_{256\times 256}$	
	Acc.	Dim.	Acc.	Dim.
$\mathcal{P}_0$	39.92	$\mathbb{R}^{1\times 250}$	37.44	$\mathbb{R}^{1\times 250}$
$\mathcal{P}_1$	50.83	$\mathbb{R}^{1\times 1000}$	50.58	$\mathbb{R}^{1\times 1000}$
$\mathcal{P}_2$	59.34	$\mathbb{R}^{1\times 4000}$	59.75	$\mathbb{R}^{1\times 4000}$
$\mathcal{P}_3$	**64.79**	$\mathbb{R}^{1\times 16000}$	63.55	$\mathbb{R}^{1\times 16000}$

Table 5.5 Fused performance (in %) of the proposed FERS.

$< \mathcal{P}_0,\mathcal{P}_1 >$	$< \mathcal{P}_0,\mathcal{P}_2 >$	$< \mathcal{P}_0,\mathcal{P}_3 >$
52.32	61.57	64.96
$< \mathcal{P}_1,\mathcal{P}_2 >$	$< \mathcal{P}_1,\mathcal{P}_3 >$	$< \mathcal{P}_2,\mathcal{P}_3 >$
59.84	63.31	**66.69**
$< \mathcal{P}_0,\mathcal{P}_1,\mathcal{P}_2 >$	$< \mathcal{P}_0,\mathcal{P}_1,\mathcal{P}_3 >$	$< \mathcal{P}_1,\mathcal{P}_2,\mathcal{P}_3 >$
59.75	61.32	63.47
$< \mathcal{P}_0,\mathcal{P}_1,\mathcal{P}_2,\mathcal{P}_3 >$		
66.61		

To compare the performance of the proposed methodology, we have computed Vgg16 [49], Gabor [50], LBP [51], HoG [52], GLCM [53], and Zernik [54] feature vectors for the employed database from the respective papers and obtain the performance in the same experimental setup as used by the proposed system. The performance due to these feature vectors are shown in Table 5.6 which shows the superiority of the proposed system.

Hence we have obtained the following findings: (i) We compute local features compared to global features. Because in the case of global feature extraction in the multi-pose and multi-level scheme, performance accuracy is not very good at all as the co-relations between pixels are less compared with local feature extraction. But, if we fragment image into smaller patches

Table 5.6 Performance comparison (in %) of the proposed FERS with the other competing methods.

Method [Ref.]	Acc.(%)
Vgg16 [49]	65.08
Gabor [50]	53.18
LBP [51]	49.34
HoG [52]	58.29
GLCM [53]	41.67
Zernik [54]	52.39
Proposed	**66.69**

then the local feature extraction technique would not give better accuracy in case of facial feature extraction as the variation between two consecutive local patches are not so high and that will not provide more discriminant information. (ii) Since the facial region contains eyes, nose, leaf and forehead region and these are the active units that may have various texture information whereas the regions other than these action areas i.e., chicks may not contain much more texture pattern. (iii) Dividing the face region into several distinct blocks can contribute more and more discriminating features but experimentally it has also been observed that these dividing the image into more and more smaller blocks, may not always contribute better texture information.

5.5 Conclusion

In this paper we have proposed a multi-level and multi-pose facial expression recognition system which has three components. In the first component, image preprocessing task has been performed where from a body silhouette image, the face region has been extracted using the facial landmark points. Then in the second component, from the extracted face region, the feature extraction techniques have been performed where various texture features like LBP, HoG, SIFT, and BoW model using dense SIFT, have been computed in the multilevel texture analysis approaches. Finally, in the last component, the SVM with linear kernel has been employed to do classification task on the computed feature vectors. Finally, it has been observed that BoW model has obtained better performance and the scores due to this feature representation scheme, are fused together to obtain better performance. The performance of the proposed system has been compared with the state-of-the-art methods using KDEF database and the comparison shows the superiority of the proposed system.

References

[1] Marcus Vinicius Zavarez, Rodrigo F Berriel, and Thiago Oliveira-Santos. Cross-database facial expression recognition based on fine-tuned deep convolutional network. In *2017 30th SIBGRAPI Conference on Graphics, Patterns and Images (SIBGRAPI)*, pages 405–412. IEEE, 2017.

[2] Christian Szegedy, Vincent Vanhoucke, Sergey Ioffe, Jon Shlens, and Zbigniew Wojna. Rethinking the inception architecture for computer vision. In *Proceedings of the IEEE conference on computer vision and pattern recognition*, pages 2818–2826, 2016.

[3] Ciprian Adrian Corneanu, Marc Oliu Simón, Jeffrey F Cohn, and Sergio Escalera Guerrero. Survey on rgb, 3d, thermal, and multimodal approaches for facial expression recognition: History, trends, and affect-related applications. *IEEE transactions on pattern analysis and machine intelligence*, 38(8):1548–1568, 2016.

[4] Albert Mehrabian and Martin Williams. Nonverbal concomitants of perceived and intended persuasiveness. *Journal of Personality and Social psychology*, 13(1):37, 1969.

[5] James A Russell and José Miguel Fernández Dols. *The psychology of facial expression*, volume 10. Cambridge university press Cambridge, 1997.

[6] Carmen Bisogni, Aniello Castiglione, Sanoar Hossain, Fabio Narducci, and Saiyed Umer. Impact of deep learning approaches on facial expression recognition in healthcare industries. *IEEE Transactions on Industrial Informatics*, 2022.

[7] Anay Ghosh, Saiyed Umer, Muhammad Khurram Khan, Ranjeet Kumar Rout, and Bibhas Chandra Dhara. Smart sentiment analysis system for pain detection using cutting edge techniques in a smart healthcare framework. *Cluster Computing*, pages 1–17, 2022.

[8] Maja Pantic and Leon J. M. Rothkrantz. Automatic analysis of facial expressions: The state of the art. *IEEE Transactions on pattern analysis and machine intelligence*, 22(12):1424–1445, 2000.

[9] Le An, Songfan Yang, and Bir Bhanu. Efficient smile detection by extreme learning machine. *Neurocomputing*, 149:354–363, 2015.

[10] Xijian Fan and Tardi Tjahjadi. Fusing dynamic deep learned features and handcrafted features for facial expression recognition. *Journal of Visual Communication and Image Representation*, 65:102659, 2019.

[11] Shengcai Liao, Anil K Jain, and Stan Z Li. Partial face recognition: Alignment-free approach. *IEEE Transactions on pattern analysis and machine intelligence*, 35(5):1193–1205, 2012.

[12] Matthew Turk and Alex Pentland. Eigenfaces for recognition. *Journal of cognitive neuroscience*, 3(1):71–86, 1991.

[13] Topi Mäenpää. *The local binary pattern approach to texture analysis: extensions and applications*. Oulun yliopisto Oulu, 2003.

[14] Milan Sonka, Vaclav Hlavac, and Roger Boyle. *Image processing, analysis, and machine vision*, volume 9. Cengage Learning, 2014.

[15] Shan Li and Weihong Deng. Deep facial expression recognition: A survey. *IEEE Transactions on Affective Computing*, 2020.

[16] H Sebastian Seung and Daniel D Lee. The manifold ways of perception. *science*, 290(5500):2268–2269, 2000.

[17] Kresimir Delac, Mislav Grgic, and Sonja Grgic. Independent comparative study of pca, ica, and lda on the feret data set. *International Journal of Imaging Systems and Technology*, 15(5):252–260, 2005.

[18] Xiaofei He, Shuicheng Yan, Yuxiao Hu, Partha Niyogi, and Hong-Jiang Zhang. Face recognition using laplacianfaces. *IEEE transactions on pattern analysis and machine intelligence*, 27(3):328–340, 2005.

[19] Meng Yang and Lei Zhang. Gabor feature based sparse representation for face recognition with gabor occlusion dictionary. In *European conference on computer vision*, pages 448–461. Springer, 2010.

[20] Matti Pietikäinen. Local binary patterns. *Scholarpedia*, 5(3):9775, 2010.

[21] Navneet Dalal and Bill Triggs. Histograms of oriented gradients for human detection. 2005.

[22] David G Lowe. Method and apparatus for identifying scale invariant features in an image and use of same for locating an object in an image, March 23 2004. US Patent 6,711,293.

[23] Ridhi Jindal and Sonia Vatta. Sift: Scale invariant feature transform. *IJARIIT*, 1:1–5, 2010.

[24] Thomas Serre, Minjoon Kouh, Charles Cadieu, Ulf Knoblich, Gabriel Kreiman, and Tomaso Poggio. A theory of object recognition: computations and circuits in the feedforward path of the ventral stream in primate visual cortex. 2005.

[25] Timo Ojala, Matti Pietikäinen, and David Harwood. A comparative study of texture measures with classification based on featured distributions. *Pattern recognition*, 29(1):51–59, 1996.

[26] Saiyed Umer, Ranjeet Kumar Rout, Chiara Pero, and Michele Nappi. Facial expression recognition with trade-offs between data augmentation and deep learning features. *Journal of Ambient Intelligence and Humanized Computing*, pages 1–15, 2021.

[27] Sanoar Hossain, Saiyed Umer, Vijayan Asari, and Ranjeet Kumar Rout. A unified framework of deep learning-based facial expression recognition system for diversified applications. *Applied Sciences*, 11(19):9174, 2021.

[28] Ligang Zhang, Dian Tjondronegoro, and Vinod Chandran. Discovering the best feature extraction and selection algorithms for spontaneous facial expression recognition. In *2012 IEEE International Conference on Multimedia and Expo*, pages 1027–1032. IEEE, 2012.

[29] Qiyu Rao, Xing Qu, Qirong Mao, and Yongzhao Zhan. Multi-pose facial expression recognition based on surf boosting. In *2015 International Conference on Affective Computing and Intelligent Interaction (ACII)*, pages 630–635. IEEE, 2015.

[30] Srinivas Gutta, Harry Wechsler, and P Jonathon Phillips. Gender and ethnic classification of face images. In *Proceedings Third IEEE International Conference on Automatic Face and Gesture Recognition*, pages 194–199. IEEE, 1998.

[31] Guangpeng Zhang and Yunhong Wang. Multimodal 2d and 3d facial ethnicity classification. In *2009 Fifth International Conference on Image and Graphics*, pages 928–932. IEEE, 2009.

[32] Zhengyou Zhang, Michael Lyons, Michael Schuster, and Shigeru Akamatsu. Comparison between geometry-based and gabor-wavelets-based facial expression recognition using multi-layer perceptron. In *Automatic Face and Gesture Recognition, 1998. Proceedings. Third IEEE International Conference on*, pages 454–459. IEEE, 1998.

[33] Marian Stewart Bartlett, Gwen Littlewort, Mark Frank, Claudia Lainscsek, Ian Fasel, and Javier Movellan. Recognizing facial expression: machine learning and application to spontaneous behavior. In *Computer Vision and Pattern Recognition, 2005. CVPR 2005. IEEE Computer Society Conference on*, volume 2, pages 568–573. IEEE, 2005.

[34] Thai Son Ly, Nhu-Tai Do, Soo-Hyung Kim, Hyung-Jeong Yang, and Guee-Sang Lee. A novel 2d and 3d multimodal approach for in-the-wild facial expression recognition. *Image and Vision Computing*, 92:103817, 2019.

[35] Nectarios Rose. Facial expression classification using gabor and log-gabor filters. In *Automatic Face and Gesture Recognition, 2006. FGR 2006. 7th International Conference on*, pages 346–350. IEEE, 2006.

[36] Tingfan Wu, Marian S Bartlett, and Javier R Movellan. Facial expression recognition using gabor motion energy filters. In *Computer Vision and Pattern Recognition Workshops (CVPRW), 2010 IEEE Computer Society Conference on*, pages 42–47. IEEE, 2010.

[37] Seyedehsamaneh Shojaeilangari, Yau Wei Yun, and Teoh Eam Khwang. Person independent facial expression analysis using gabor features

and genetic algorithm. In *Information, Communications and Signal Processing (ICICS) 2011 8th International Conference on*, pages 1–5. IEEE, 2011.

[38] Wenfei Gu, Cheng Xiang, YV Venkatesh, Dong Huang, and Hai Lin. Facial expression recognition using radial encoding of local gabor features and classifier synthesis. *Pattern recognition*, 45(1):80–91, 2012.

[39] Huiting Wu, Yanshen Liu, Yi Liu, and Sanya Liu. Efficient facial expression recognition via convolution neural network and infrared imaging technology. *Infrared Physics & Technology*, 102:103031, 2019.

[40] Timur R Almaev and Michel F Valstar. Local gabor binary patterns from three orthogonal planes for automatic facial expression recognition. In *2013 Humaine Association Conference on Affective Computing and Intelligent Interaction*, pages 356–361. IEEE, 2013.

[41] Guoying Zhao and Matti Pietikainen. Dynamic texture recognition using local binary patterns with an application to facial expressions. *IEEE transactions on pattern analysis and machine intelligence*, 29(6):915–928, 2007.

[42] Xusheng Wang, Xing Chen, and Congjun Cao. Human emotion recognition by optimally fusing facial expression and speech feature. *Signal Processing: Image Communication*, page 115831, 2020.

[43] Ruicong Zhi, Caixia Zhou, Tingting Li, Shuai Liu, and Yi Jin. Action unit analysis enhanced facial expression recognition by deep neural network evolution. *Neurocomputing*, 2020.

[44] Yuqian Zhou and Bertram E Shi. Action unit selective feature maps in deep networks for facial expression recognition. In *2017 International Joint Conference on Neural Networks (IJCNN)*, pages 2031–2038. IEEE, 2017.

[45] Yanpeng Liu, Yuwen Cao, Yibin Li, Ming Liu, Rui Song, Yafang Wang, Zhigang Xu, and Xin Ma. Facial expression recognition with pca and lbp features extracting from active facial patches. In *Real-time Computing and Robotics (RCAR), IEEE International Conference on*, pages 368–373. IEEE, 2016.

[46] David G Lowe. Distinctive image features from scale-invariant keypoints. *International journal of computer vision*, 60(2):91–110, 2004.

[47] Mihran Tuceryan, Anil K Jain, et al. Texture analysis, handbook of pattern recognition & computer vision, 1993.

[48] Yin Zhang, Rong Jin, and Zhi-Hua Zhou. Understanding bag-of-words model: a statistical framework. *International Journal of Machine Learning and Cybernetics*, 1(1-4):43–52, 2010.

[49] Karen Simonyan and Andrew Zisserman. Very deep convolutional networks for large-scale image recognition. *arXiv preprint arXiv:1409.1556*, 2014.

[50] Jun Ou, Xiao-Bo Bai, Yun Pei, Liang Ma, and Wei Liu. Automatic facial expression recognition using gabor filter and expression analysis. In *2010 Second International Conference on Computer Modeling and Simulation*, volume 2, pages 215–218. IEEE, 2010.

[51] Caifeng Shan, Shaogang Gong, and Peter W McOwan. Robust facial expression recognition using local binary patterns. In *IEEE International Conference on Image Processing 2005*, volume 2, pages II–370. IEEE, 2005.

[52] Pranav Kumar, SL Happy, and Aurobinda Routray. A real-time robust facial expression recognition system using hog features. In *2016 International Conference on Computing, Analytics and Security Trends (CAST)*, pages 289–293. IEEE, 2016.

[53] Gorti SatyanarayanaMurty, J SasiKiran, and V Vijaya Kumar. Facial expression recognition based on features derived from the distinct lbp and glcm. *International Journal of Image, Graphics and Signal Processing*, 2(1):68–77, 2014.

[54] Mohammed Saaidia, Narima Zermi, and Messaoud Ramdani. Facial expression recognition using neural network trained with zernike moments. In *2014 4th International Conference on Artificial Intelligence with Applications in Engineering and Technology*, pages 187–192. IEEE, 2014.

6

A Texture Features-based Method to Detect the Face Spoofing

Somenath Dhibar and Bibhas Chandra Dhara

Department of Information Technology, Jadavpur University, Kolkata, India
E-mail: somenath.ju@gmail.com; bcdhara@gmail.com

Abstract

Biometric authentication is very important nowadays. This authentication system has various attacking points that are vulnerable to different attacks. Although the biometric traits (face, iris, voice, etc.) seem personal for individuals, though unknowingly, we disclose those traits publicly in our daily activities. Facebook, Twitter, WhatsApp and Instagram are the most frequently used social networking sites. These are used to share similar personal information and make it easily available to the public. Therefore, a person can easily get biometric traits of others and may misuse this or can gain illegitimate access and advantages. To secure this, a spoofing detection algorithm is needed. The present work can detect whether the face presented before a camera is real or fake. Our proposed system consists of three phases: the phase is preprocessing, where the face region is marked and cropped using facial landmark points. The second phase is feature extraction where we compute the dense scale invariant feature transform (DSIFT) descriptors from the detected facial region and represent them using three different approaches. The last phase is for classification, where SVM is used for the classification task and it determines whether the image is an original face or a fake. We have tested the performance of the proposed work using the NUAA database, which is publicly available. The accuracy of the proposed system is superior to some existing methods.

Keywords: Spoofing detection, face spoof detection, Dsift feature, SVM classifier.

6.1 Introduction

Biometric authentication is the most significant task to maintain the security of a system. It is widely used by more and more people and it is significantly progressing because of its excellent performance. Due to the flexibility of this authentication system, it has continued its upward trends.

Several authentication techniques are already in the market [1], such as passwords [2], PINs [3], smart cards [4], tokens [5], keys [6] etc. But, all these are either easy to forget, or they can be lost or stolen by others. However, it is impossible to lose an individual's biological traits, neither can it be forgotten nor stolen by anyone.

Face recognition is one of the fastest biometric authentication systems as compared with other biometric traits such as fingerprint and iris recognition. For example, in a surveillance system, face recognition gives highly reliable authentication as there is no need to put their hands on radar (fingerprint) or carefully position their eyes in front of a scanner (iris recognition). Millions of people are using different biometric traits to secure their highly sensitive information.

The 2D face authentication system uses 2D faces to identify the individuals [7]. The challenging part of 2D face recognition are viewpoints, occlusion, brightness, and contrast of outdoor lights [8].

Unfortunately, biometric traits are easily available in the public domain (social networking sites). Since individuals' face images can possibly be publicly available on social media or any other source. So, hiding this data from society is very challenging and sometimes not possible at all. Those images can be printed on paper, can be used to make sophisticated masks or can help to generate a 3D model of the targeted person by presenting those in front of the camera [9]. To make the biometric system more secure, we need to look into another possible way without hiding the biometric traits from the outdoors. The gaining of illegitimate access and advantages by masquerading as someone else can be defined as spoofing in biometric systems and spoofing of a face recognition system can be called face spoofing.

The vulnerability of a face biometric system due to spoofing attacks is mostly overlooked. The National Institute of Standards and Technology (NIST) in the United States has listed some weaknesses in the National Vulnerability Database [10]. Inspired by image quality assessment [11], a face spoofing detection technique is proposed where differences between the reflection of light and characterization of printing artifacts have been considered. The spoofing detection techniques can enable people to access their secure data.

In this article, we have presented an algorithm for face spoofing detection. The center of attention of this chapter is to encapsulate how an analysis of an image can secure it from another copy of the same image but a spoof one. To avail this, we have employed three steps: (i) preprocessing (ii) feature extraction, and (iii) classification. The proposed approach scrutinizes the texture information of face images using the dense scale invariant feature transform (DSIFT) and encodes the texture feature. The features are taken to the support vector machine (SVM) [10] classifier. The role of SVM is to determine whether a live person is present in front of the camera or a spoofed image is presented. In our experiment, we have used a publicly available database (NUAA) [12] that contains genuine and imposter photographs. Our proposed architecture can verify that the results of face spoofing recognition using local features over global features can offer significantly better performance.

This chapter is organized as follows: Section 6.2 describes the previous works. Section 6.3 presents the proposed method. The experimental results and discussion are given in Section 6.4. In Section 6.5 the conclusions have been made.

6.2 Literature Review

The face recognition task consists of some sub-tasks in a series: face detection, feature extraction, and classification. The block diagram of the proposed face recognition/verification is shown in Figure 6.1. The various techniques of face recognition have been highlighted in [13–16]. They can be classified as geometrically dependent local feature-based technique, holistic template matching technique, and the hybrid system.

Non-linearity is the most challenging part of an automatic face recognition system and also very hard to solve. This problem can be seen as a template matching problem in a high dimensional space. In high dimensional space more complex computation is needed to find the possible match. Therefore, to reduce this problem the dimension should be reduced into a low dimensional space. For this, a dimensionality reduction technique is needed. Eigenfaces [17], principal component analysis (PCA) [18, 19, 20] and linear discriminant analysis (LDA) [21] are widely used to reduce the dimension of the feature space. Some researchers have combined the PCA and LDA [22, 23]. Some hybrid versions of LDA like DLDA (Direct LDA) and FLDA (Fractional LDA) have been proposed in [24], where too-closed output classes are avoided using weighted functions.

The elastic bunch graph matching (EBGM) [25] is another technique that has excellent performance, but to perform better it requires large images. In contrast to the verification task, the subspace LDA system [26] has a very good result and also works for both large and small images. Thus, for a specific task, one should choose the scheme carefully. Although many systems have achieved brilliant success, in most cases two prominent issues need to be addressed. They are illumination and pose problems, i.e., change of the projection vector. Many face recognition systems have adopted the methods to handle these issues [27–30]. So, it becomes very difficult to recognize the individuals in an uncontrolled environment (i.e., in surveillance video clips). In an article [13], a 3D model is proposed to amplify the existing 2D approach. Further development has been done in [31] with discriminant common vectors (DCV).

The main disadvantage of the above approaches is their linearity. Neural network (NN) [32] is one of the widely accepted nonlinear solutions to the face recognition system. Bai *et al.* [33] have worked on the detection of spoofing attacks, where they have extracted micro-texture features from the distinct component of a recaptured image. Finally, classify them using a linear SVM classifier. The major disadvantage of the work is that the input must be a high-resolution image; otherwise, the system fails to classify the images. Currently, deep learning-based methods are gaining attraction in spoofing detection. Some researchers have used it in facial expression recognition [34, 35]. Some frameworks are also established for smart healthcare [35] and facial expression recognition-based systems for diverse applications [37]. Gao *et al.* [38] have introduced another interesting approach. The LeNet-5 CNN structure can detect the movement of the human eye blinking as proposed by [39] and using this a dynamic biometric authentication system (DBAS) is proposed in [40]. A combined analysis of entropy and quality-based research work [41] shows a greater performance over the texture analysis.

The biometric spoofing attack can be done in many ways. Video replay attacks [42], 3D mask attacks [43], and printing attacks [44, 45] are very common for face spoofing. Zhang *et. al.* [46], proposed spoofing detection algorithms for the face, which are based on the color texture Markov feature

(CTMF). In this work, classification and dimension reduction tasks have been materialized by using SVM recursive feature elimination (SVM-RFE), where the co-located pixels value between the real face and the spoofed one are analyzed carefully and observe the differences between them. These texture differences are captured by a directional differential filter and then the Markov model is used to define the features. Finally, SVM-RFE is used for spoofing detection. Yang *et al.* [47, 48] have proposed a novel spatio-temporal mechanism for spoofing detection. The authors have used global features for the temporal information and local features for spatial information and combined these features to detect the spoofing attack. This spatio-temporal anti-spoofing network (STASN) model has three parts. The first part is CNN-LSTM, which predicts the binary classified output. The second part learns the offset from the CNN features and detects the participating regions of the sequence images. This region is then fed to the last part. Then fused it for prediction. According to the author, along with a good network, we need good data, which is also very important for face anti-spoofing algorithms. Liu *et al.* [49] have proposed a novel deep tree network (DTN). They define the unknown spoofing attack as zero-shot facial anti-spoofing (ZSFA). This model is trained in an unsupervised way for better analysis of ZSFA. The authors also have created a database, SiW-M. Shao *et al.* [50] proposed a face anti-spoofing method that has the generalization ability. Daniel *et al.* [51] redefined the generalization ability and proposed a novel face spoofing detection technique using depth-metric learning.

6.3 Proposed Methodology

Spoofing can be like playing a video of a person instead of the live person or any publicly available image of the person taken from other sources (faceprints). These faceprints usually have some printing quality defects. If we analyze the texture of it, that can be easily recognized. Let's say, we consider a picture that has been taken from another picture and analyze the reflection of lights from that picture. We can detect the differences, because the face is a 3D object and it is complex and non-rigid whereas the picture is a planer rigid object. The proposed face spoofing detection technique has three subsequent phases: (i) preprocessing (to detect the facial region), (ii) feature extraction, and (iii) classification. The block diagram of the proposed method is shown in Figure 6.1. The phases of the proposed method are presented in the following sub-sections.

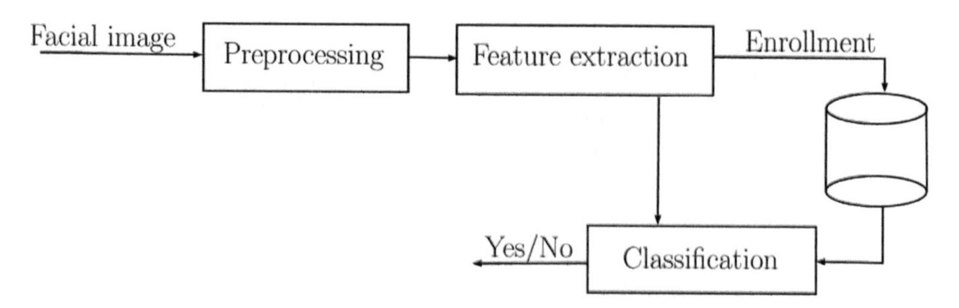

Figure 6.1 Block diagram of the proposed face spoofing detection.

6.3.1 Preprocessing

In an arbitrary environment, handling a face biometric system is greatly challenging because of the wide variety of scenarios. The major demanding part of a face recognition system is the face detection and detection of facial landmarks [52]. In the present work, the preprocessing step is used to detect the facial region and facial landmarks. To solve this problem, in this work, we have adopted the tree structure model as proposed in [53]. This tree model is designed using mixtures of the trees along with a pool of landmark regions. This model is then applied to the input face image and obtains certain landmarks. The identified landmarks help to determine the facial region. The result of the preprocessing step is shown in Figure 6.2.

6.3.2 Feature extraction

It is observed that every individual (even in the case of twins) has a unique and distinct facial biometric trait. The texture description is one of the techniques to represent the face. In this work, we have used both structural and statistical textures for this purpose.

In the structural approach, the frequency of occurrences of certain patterns has been considered. The statistical analysis of the texture is based on the spatial layout of the intensity pattern within a certain region. Among statistical approaches, in this work, the edginess of texture (EoT) [54] has been adopted. The EoT of a region is defined as the number of edge points per unit area, i.e., EoT defines the intricacy of texture. This EoT gives an excellent measure due to its distinctiveness and robustness properties, if this is computed over a small patch.

The concept of EoT has been adopted in different texture features like scale invariant feature transform (SIFT) [55, 56], Dense SIFT (DSIFT) [57],

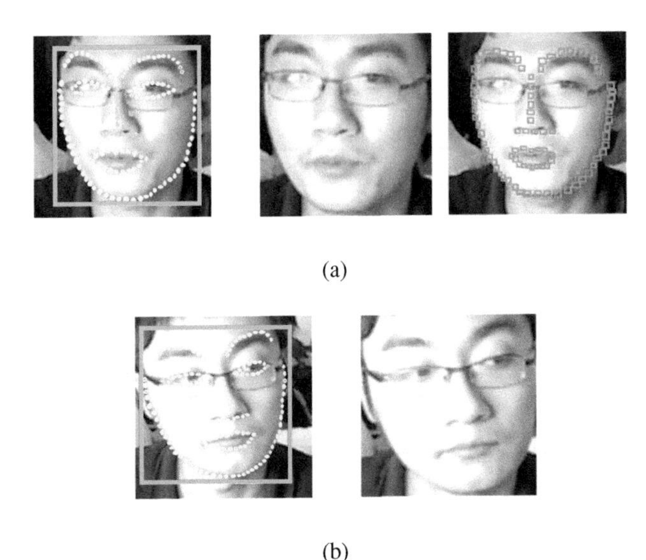

Figure 6.2 Illustration of face preprocessing technique: (a) Frontal face image, (b) Profile face image.

Daisy descriptor (DD) [58], gradient local auto correlation (GLAC) [59], histogram of oriented gradient (HOG) [60, 61], speeded up robust features (SURF) [62]. Individually, these features can be extracted from the detected face region. In the present work, we have used the DSIFT features to detect the spoofing of facial images.

Here, we have used the DSIFT feature because the descriptors are invariant to the geometric affine transform, many features can be obtained using the dense SIFT algorithms. A number of redundant information can be captured from an image by using this DSIFT. Whereas only relevant information can be found using the normal SIFT algorithms. One must say that a lot of irrelevant data will be there if we use DSIFT, but this unrelated information can be removed in the training phase and this will give better performance. The performance (see Section 6.4) of the proposed system supports this. Using DSIFT, we can also get an overall better feature description with additional characteristics of the flatter regions.

In order to classify objects efficiently, it is obvious that global features have less impact. So, in this work, instead of the global features, we have concentrated on the local features. Hence, we have considered the small patches of size $p \times p$ around the landmarks. From each patch, we have computed $v_i = (x_{i1}, x_{i2}, \cdots, x_{it})^T \in R^{t \times D}$ where t is the number of the

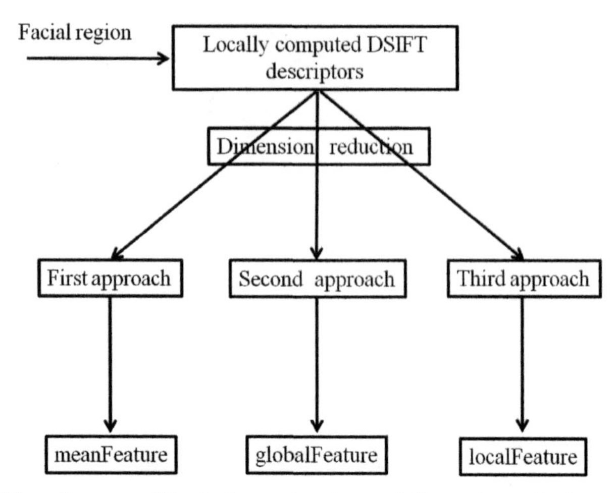

Figure 6.3 Flow diagram of the feature extraction and dimension reduction approaches.

descriptors in each patch and dimension of each descriptor is D, i.e., $x_{ij} = (x_{ij}(1), x_{ij}(2), \cdots, x_{ij}(D))$. For the facial image, we have obtained n patches as returned from the tree structure model in the preprocessing step of the proposed model. From n patches, we have a collection of descriptors $V = (v_1, v_2, \cdots, v_n)^T$. Therefore, the image can be described by

$$V = (x_{11}, x_{12}, \cdots, x_{1t}, x_{21}, x_{22}, \cdots, x_{2t}, \cdots, x_{n1}, x_{n2}, \cdots, x_{nt})^T \in R^{nt \times D}$$

The dimension of the descriptor of the facial image, I, is very high and that is not suitable in real-time applications. Hence, we need to reduce the dimension of the descriptor of the I. In the present work, we have proposed three different approaches that are applied to reduce the dimension of the descriptor. The flow diagram of the feature extraction and reduction is given in Figure 6.3. Accordingly, we have three approaches for spoof detection. In the following, we have presented the methods that are used to reduce the dimension.

First approach: In this technique each patch is represented by the mean descriptor, $mean_{v_i}$, where

$$mean_{v_i}(k) = \frac{1}{t} \sum_{j=1}^{t} x_{ij}(k) \quad \text{for } k = 1, 2, \cdots, D$$

The algorithmic sketch to compute the mean descriptor ($mean_{v_i}$) of the facial image is given in **Algorithm 1:** $meanFeature$().

Algorithm 1: $meanFeature(I, L = \{l_1, l_2, \cdots, l_n\})$

Input: facial image I, thc locations of the landmarks L

Output: mean feature vector $meanFeature_I$

1. For each landmark position l_i

 1.1 Compute descriptor $v_i = (x_{i1}, x_{i2}, \cdots, x_{it})^T$

 1.2 Compute mean descriptor

  1.2.a. For k = 1 to D

  1.2.a.i $mean_{v_i}(k) = \frac{1}{t} \sum_{j=1}^{t} x_{ij}(k)$

2. $meanFeature_I(i) \leftarrow mean_{v_i}$

Second approach: In the first approach, to represent a face, the size of the feature vector is Dt, which is also significantly large. In the second approach, to further reduce the size of the descriptor, we have adopted the concept of the bag of words (or vector quantization) where a predefined codebook is used. The codebook can be designed using some clustering algorithms like LBG or K-means algorithm. For each face image, we have computed the descriptors and for each descriptor, we determine the closest codeword and accordingly frequency of the descriptors, with respect to the codebook, is computed. This frequency has normalized and each image is defined by a vector of size K. Where K is the number of codewords in the codebook. This frequency distribution gives a global textural description of the given image I, as we consider the average of these features. Let us refer to the computed feature vector as $globalFeature_I$. The algorithm which computes this feature vector of a given face image is presented in **Algorithm 2:** globalFeature() and the function 'matching()' gives the best matched codeword for the feature vector x_{ir} from the codebook CB.

Algorithm 2: $gloablFeature(I, L = \{l_1, l_2, \cdots, l_n\}, CB)$

Input:facial image I, the locations of the landmarks L, codebook CB

Output: feature vector $globalFeature_I$

1. $globalFeature_I(1 \cdots K) \leftarrow 0$

2. For each landmark position l_i

   2.1 Compute descriptor $v_i = (x_{i1}, x_{i2}, \cdots, x_{it})^T$

   2.2 For r = 1 to t // each descriptor

   2.2.a. $p = matching(CB, x_{ir})$

   2.2.a.i $globalFeature_I(p) = globalFeature_I(p) + 1$

3. $globalFeature_I = \frac{globalFeature_I}{n}$

Third approach: In the second approach, we have used the global codebook and hence the corresponding feature is global in some sense. But, the major drawback of the second approach is the dimension of the feature vectors. In the third approach, we follow a similar approach as the second approach. But, here size of the codebook (k) is much smaller (i.e., $k << K$) and with respect to each descriptor of the image I the frequency distribution has computed, and then these are concatenated to represent the image I. Since for each descriptor, we have a separate distribution, this feature vector contains more local information. Let us refer to the feature vector obtained using this approach as $localFeature_I$. The algorithmic sketch of this approach is **Algorithm 3:** localFeature() (same as **Algorithm 2**).

Algorithm 3: $localFeature(I, L = \{l_1, l_2, \cdots, l_n\}, CB)$

Input:facial image I, the locations of the landmarks L, codebook CB

Output: feature vector $localFeature_I$

---

1. $localFeature_I = \Phi$

2. For each landmark position l_i

   2.1 $feature_I\ (1 \cdots k) \leftarrow 0$

   2.2 Compute descriptor $v_i = (x_{i1}, x_{i2}, \cdots, x_{it})^T$

   2.2 For r = 1 to t // each descriptor

   2.2.a. p = matching(CB, x_{ir})

   2.2.a.i $feature_I(p) = feature_I(p) + 1$

2.3 $feature_I = \frac{feature_I}{t}$

3. $localFeature_I = localFeature_I + feature_I$ // '+' $\rightarrow$ concatenation

6.3.3 Classification

In this work, to classify the input image as an original image or spoofed image, we have used a nonlinear SVM classifier with a radial basis function kernel. First, the SVM classifier is trained using a set of positive (real faces) and negative (spoofed faces) samples. For authentication, the test sample, different from the training samples, have been used to test the efficiency of the proposed approaches. From the above, we note that three approaches have been used to reduce the dimension of the feature vectors. Hence, we have designed three different models to design the authentication system. The proposed models are named as $meanFeature$, $globalFeature$, and $localFeature$ according to the feature reduction approaches.

6.4 Experimental Results

This work is implemented in the MATLAB environment on a system having Windows 10 Pro as an OS. It has an Intel Core I5 processor with a speed of 2.40 GHz and 8 GB RAM.

We have tested the performance of the proposed work using the NUAA photograph imposter database [7], which is a publicly available database. This database contains both real client accesses and fake one for photo attacks. The photographs have been taken manually in duration of two weeks and three sessions each day. In addition, while taking the images, different conditions like illumination conditions and environmental changes are taken care of. Some images from the NUAA database are shown in Figure 6.4. The genuine images are shown in the first row and the imposter images are shown in the second row. In the database, two types of images are recorded; one is for client accesses and another one for spoofing attacks. Normal webcams have been used to capture these images and the resolution of the images is 640×480 pixels. Here, each subject has 500 sample images. While capturing the images, special care was taken so that the person looks as static as possible and eye-blinking is as minimal as possible. This care is to minimize the difference between the live and spoofed images. In contrast, while capturing the spoofed photographs, five distinct graphic photograph attacks were imitated with varying motions using the 2D face prints. This data

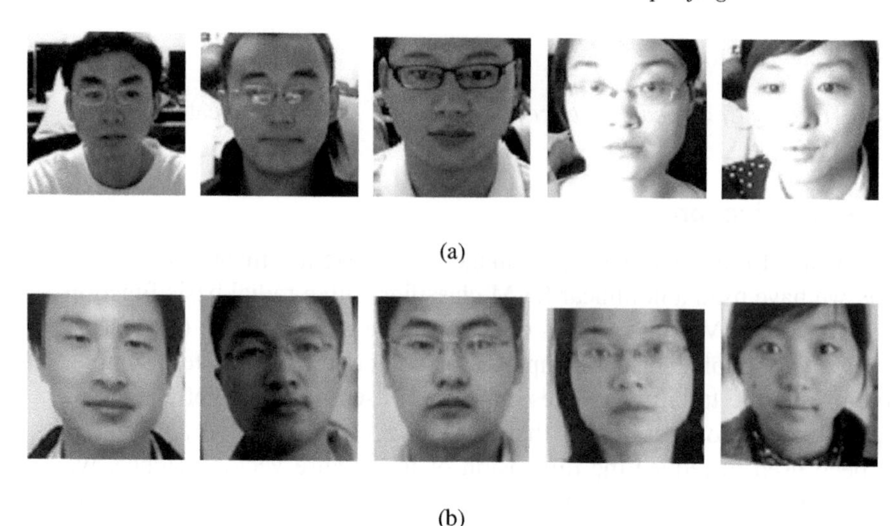

(a)

(b)

Figure 6.4 Sample test images from the NUAA database: (a) Genuine images, (b) Imposter images.

set consists of 15 different subjects on real face images and spoofed images with different qualities, like photograph quality and laser-quality prints.

The dataset of the database is separated into two different sets. One is used for training purposes (called as training set) and another one is for testing purposes (called as test set). So, the training set and test set are exclusive, i.e., there is no common image.

The training dataset includes nine real clients and from these subjects altogether 1743 real face images have been used for training purposes. Among those training images 889 images were captured in the first session and 854 images were taken in the second session. Likewise, thc training dataset for the imposter subject consists of the same nine clients and a total of 1748 imposter images have been considered, where 855 images were generated in the first session and 893 images were captured in the second session.

The test dataset includes the images of the database which are not used for the training purposes (i.e., the training images of the six subjects are not included in the test set). The test set consists of images (may real or spoofed) of nine subjects. From these nine subjects, there are three subjects from the training set but training images are excluded. There are six clients who have not participated in the training set, the images of all those clients are also included in the test set. The test set has 3362 client samples from the genuine images and 5761 imposter images. In this implementation, all the face images are normalized into 200×200 pixels. Some samples of the genuine and imposter images of the training set are shown in Figure 6.5.

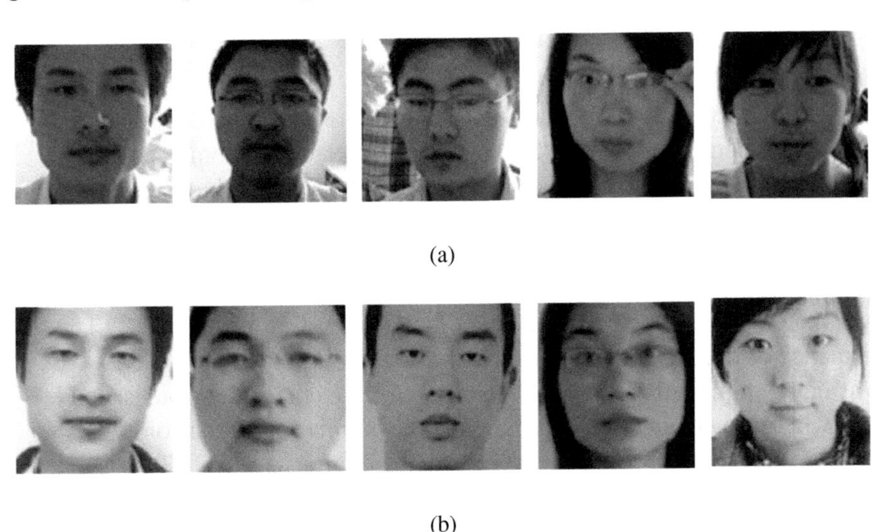

Figure 6.5 Training sample images: (a) Genuine images, (b) Imposter images.

The summary of the training and test sets are given in the Table 6.1.

Table 6.1 Summary of training and test datasets in this experiment.

	Genuine images	Imposter images
TEST	3362	5761
TRAIN	1743	1748

6.4.1 Results and discussion

In this section, we have reported the result of our proposed models and then we compared the performance with the existing methods. For the comparison purpose, we have considered the articles which have used the NUAA dataset and we have quoted the result from those respective articles.

During face preprocessing as discussed earlier, a tree-structured model has been used. That model has computed some landmark points for every face (frontal and profile faces). These landmark points have been used to extract the region of the face and after detecting the region of the face, the image is

Table 6.2 Performance of the proposed system using proposed three models.

Method used	Accuracy (%)
meanFeature	72.07
globalFeature	58.24
localFeature	78.15

normalized into 200×200. We have used a patch w of size 50×50 and the total face has been partitioned into n = 16 distinct patches. From each patch, we have computed descriptors with dimension 128. Then we have tested the performance according to the proposed models.

In the first model, where we have computed the mean features (i.e., $meanFeature_I$) and we obtained a 16×128 dimensional descriptor. This feature vector is used for both verification and identification purposes. The performance of this model is reported in Table 6.2.

We have employed the global feature ($globalFeature$), in our second model, where we fixed a codebook with the number of codewords K = 1000. The performance of $globalFeature$ is demonstrated in Table 6.2. We observed that the performance of $meanFeature$ model is better than $globalFeature$ model.

In the third model, we have represented the local feature, that is ($localFeature$). Here, we have considered a codebook with size k = 500. The performance of the $localFeature$ model is given in Table 6.2. From this table, we noted that the $localFeature$ model outperforms the other two models. From the performance of the proposed three models of the current system, we may note the followings:

1. DSIFT descriptor for the $localFeature$ has more discerning power to analyze the texture information of the face pattern.
2. Among the $meanFeature_I$, $globalFeature_I$, and $localFeature_I$ feature vectors, $localFeature_I$ achieves an outstanding performance. The $localFeature_I$ acquires more and more minute details of local information to differentiate face patterns among the inter-class of the face images while preserving the similarity among the intra-class of the images.

Therefore, for further discussion, we continue with the local feature, i.e., $localFeature$ model of the proposed system and henceforth we refer to this model as our proposed system. Now, we study the performance of the proposed system by varying the patch size and changing the size of the codebook. This study is summarized in Table 6.3.

Table 6.3 Performance of the proposed system (i.e., $localFeature$ using various patch size and cluster center).

Patch size	Cluster center	Accuracy (%)
50×50	500	78.15
30×30	500	80.18
16×16	500	82.35
16×16	256	81.90

Table 6.4 Performance comparison of the proposed system with the existing systems.

Methods	Features	Accuracy (%)
Akbulut *et al.* [63]	CNN model	76.31
Şengür *et al.* [64]	AlexNet fc7 features	79.24
Şengür *et al.* [64]	VggNet fc7 features	81.35
Zhang *et al.* [65]	M-DoG	81.80
Proposed system	$DSIFT$ features + SVM	82.35

Observing the above table, we may conclude that if we can extract more and more local features, then the recognition system will perform better. It is observed that the performance also depends on the codebook. So the challenges are to find out the optimal patch size and a suitable codebook.

In Table 6.4, we have reported a comparative study. We note that the proposed system performs better compared to the other systems (in the case of the proposed system, we have considered the best performance, i.e., the $localFeature$ model). It may be noted that in our system we have used DSIFT features and SVM is used for classification purposes, whereas in some of the other methods advanced techniques have been used. The main disadvantage of the convolutional neural network is that, for better performance, huge input data is needed for the training phase.

6.5 Conclusion

In this article, we have proposed a face spoofing detection system using a locally computed DSIFT feature with a nonlinear SVM classifier with a radial basis function kernel as a classifier. The proposed system outperforms the existing methods. Here, we have observed that local features give better performance. In this implementation, we have used the NUAA database which has only a few samples. So, our target is to use a larger database to establish the applicability of the proposed system. We will also try to design some other local features to improve the performance of the spoof detection.

References

[1] Ignacio Velásquez, Angélica Caro, and Alfonso Rodríguez. Authentication schemes and methods: A systematic literature review. *Information and Software Technology*, 94:30–37, 2018.

[2] Mohammad Mannan and Paul C Van Oorschot. Using a personal device to strengthen password authentication from an untrusted computer. In *International Conference on Financial Cryptography and Data Security*, pages 88–103. Springer, 2007.

[3] Dhruv Kumar Yadav, Beatrice Ionascu, Sai Vamsi Krishna Ongole, Aditi Roy, and Nasir Memon. Design and analysis of shoulder surfing resistant pin based authentication mechanisms on google glass. In *International conference on financial cryptography and data security*, pages 281–297. Springer, 2015.

[4] Min-Shiang Hwang, Song-Kong Chong, and Te-Yu Chen. Dos-resistant id-based password authentication scheme using smart cards. *Journal of Systems and Software*, 83(1):163–172, 2010.

[5] Manuel Koschuch, Matthias Hudler, Hubert Eigner, and Zsolt Saffer. Token-based authentication for smartphones. In *2013 International Conference on Data Communication Networking (DCNET)*, pages 1–6. IEEE, 2013.

[6] Masood Ahmad, Ateeq Ur Rehman, Nighat Ayub, MD Alshehri, Muazzam A Khan, Abdul Hameed, and Halil Yetgin. Security, usability, and biometric authentication scheme for electronic voting using multiple keys. *International Journal of Distributed Sensor Networks*, 16(7):1550147720944025, 2020.

[7] WenYi Zhao and Rama Chellappa. Image-based face recognition: Issues and methods. *Optical engineering-New York-marcel dekker incorporated-*, 78:375–402, 2002.

[8] Andrea F Abate, Michele Nappi, Daniel Riccio, and Gabriele Sabatino. 2d and 3d face recognition: A survey. *Pattern recognition letters*, 28(14):1885–1906, 2007.

[9] Anita Babu, Vince Paul, and Dimple Elizabeth Baby. An investigation of biometric liveness detection using various techniques. In *2017*

International Conference on Inventive Systems and Control (ICISC), pages 1–5. IEEE, 2017.

[10] Jukka Määttä, Abdenour Hadid, and Matti Pietikäinen. Face spoofing detection from single images using texture and local shape analysis. *IET biometrics*, 1(1):3–10, 2012.

[11] Zhou Wang, Alan C Bovik, and Ligang Lu. Why is image quality assessment so difficult? In *2002 IEEE International conference on acoustics, speech, and signal processing*, volume 4, pages IV–3313. IEEE, 2002.

[12] Xiaoyang Tan, Yi Li, Jun Liu, and Lin Jiang. Face liveness detection from a single image with sparse low rank bilinear discriminative model. In *European Conference on Computer Vision*, pages 504–517. Springer, 2010.

[13] Wenyi Zhao, Rama Chellappa, P Jonathon Phillips, and Azriel Rosenfeld. Face recognition: A literature survey. *ACM computing surveys (CSUR)*, 35(4):399–458, 2003.

[14] Suvendu Mandal and Bibhas Chandra Dhara. A hybrid face recognition method based on structural and holistic features. In *2009 seventh international conference on advances in pattern recognition*, pages 441–444. IEEE, 2009.

[15] Srinivas Nagamalla and Bibhas Chandra Dhara. A novel face recognition method using facial landmarks. In *2009 Seventh International Conference on Advances in Pattern Recognition*, pages 445–448. IEEE, 2009.

[16] Saiyed Umer, Bibhas Chandra Dhara, and Bhabatosh Chanda. Face recognition using fusion of feature learning techniques. *Measurement*, 146:43–54, 2019.

[17] Matthew Turk and Alex Pentland. Eigenfaces for recognition. *Journal of cognitive neuroscience*, 3(1):71–86, 1991.

[18] Svante Wold, Kim Esbensen, and Paul Geladi. Principal component analysis. *Chemometrics and intelligent laboratory systems*, 2(1-3):37–52, 1987.

[19] Juwei Lu, Kostas N Plataniotis, and Anastasios N Venetsanopoulos. Regularized discriminant analysis for the small sample size problem in face recognition. *Pattern recognition letters*, 24(16):3079–3087, 2003.

[20] Aleix M Martinez and Avinash C Kak. Pca versus lda. *IEEE transactions on pattern analysis and machine intelligence*, 23(2):228–233, 2001.

[21] Suresh Balakrishnama and Aravind Ganapathiraju. Linear discriminant analysis-a brief tutorial. *Institute for Signal and information Processing*, 18(1998):1–8, 1998.

[22] Li-Fen Chen, Hong-Yuan Mark Liao, Ming-Tat Ko, Ja-Chen Lin, and Gwo-Jong Yu. A new lda-based face recognition system which can solve the small sample size problem. *Pattern recognition*, 33(10):1713–1726, 2000.

[23] Hua Yu and Jie Yang. A direct lda algorithm for high-dimensional data-with application to face recognition. *Pattern recognition*, 34(10):2067–2070, 2001.

[24] Juwei Lu, Konstantinos N Plataniotis, and Anastasios N Venetsanopoulos. Face recognition using lda-based algorithms. *IEEE Transactions on Neural networks*, 14(1):195–200, 2003.

[25] Madasu Hanmandlu, Divya Gupta, and Shantaram Vasikarla. Face recognition using elastic bunch graph matching. In *2013 IEEE Applied Imagery Pattern Recognition Workshop (AIPR)*, pages 1–7. IEEE, 2013.

[26] Wenyi Zhao, Rama Chellappa, P Jonathon Phillips, et al. *Subspace linear discriminant analysis for face recognition*. Citeseer, 1999.

[27] Wen Yi Zhao and Rama Chellappa. Illumination-insensitive face recognition using symmetric shape-from-shading. In *Proceedings IEEE Conference on Computer Vision and Pattern Recognition. CVPR 2000 (Cat. No. PR00662)*, volume 1, pages 286–293. IEEE, 2000.

[28] P Jonathon Phillips, Hyeonjoon Moon, Syed A Rizvi, and Patrick J Rauss. The feret evaluation methodology for face-recognition algorithms. *IEEE Transactions on pattern analysis and machine intelligence*, 22(10):1090–1104, 2000.

[29] Peter N. Belhumeur, Joao P Hespanha, and David J. Kriegman. Eigenfaces vs. fisherfaces: Recognition using class specific linear projection. *IEEE Transactions on pattern analysis and machine intelligence*, 19(7):711–720, 1997.

[30] Yael Adini, Yael Moses, and Shimon Ullman. Face recognition: The problem of compensating for changes in illumination direction. *IEEE Transactions on pattern analysis and machine intelligence*, 19(7):721–732, 1997.

[31] Hakan Cevikalp, Marian Neamtu, Mitch Wilkes, and Atalay Barkana. Discriminative common vectors for face recognition. *IEEE Transactions on pattern analysis and machine intelligence*, 27(1):4–13, 2005.

[32] Warren S Sarle. Neural networks and statistical models. 1994.

[33] Jiamin Bai, Tian-Tsong Ng, Xinting Gao, and Yun-Qing Shi. Is physics-based liveness detection truly possible with a single image? In *Proceedings of 2010 IEEE International Symposium on Circuits and Systems*, pages 3425–3428. IEEE, 2010.

[34] Carmen Bisogni, Aniello Castiglione, Sanoar Hossain, Fabio Narducci, and Saiyed Umer. Impact of deep learning approaches on facial expression recognition in healthcare industries. *IEEE Transactions on Industrial Informatics*, 2022.

[35] Saiyed Umer, Ranjeet Kumar Rout, Chiara Pero, and Michele Nappi. Facial expression recognition with trade-offs between data augmentation and deep learning features. *Journal of Ambient Intelligence and Humanized Computing*, pages 1–15, 2021.

[36] Anay Ghosh, Saiyed Umer, Muhammad Khurram Khan, Ranjeet Kumar Rout, and Bibhas Chandra Dhara. Smart sentiment analysis system for pain detection using cutting edge techniques in a smart healthcare framework. *Cluster Computing*, pages 1–17, 2022.

[37] Sanoar Hossain, Saiyed Umer, Vijayan Asari, and Ranjeet Kumar Rout. A unified framework of deep learning-based facial expression recognition system for diversified applications. *Applied Sciences*, 11(19):9174, 2021.

[38] Xinting Gao, Tian-Tsong Ng, Bo Qiu, and Shih-Fu Chang. Single-view recaptured image detection based on physics-based features. In *2010 IEEE International Conference on Multimedia and Expo*, pages 1469–1474. IEEE, 2010.

[39] Marwa Saied, Ayman Elshenawy, and Mohamed M Ezz. A novel approach for improving dynamic biometric authentication and verification of human using eye blinking movement. *Wireless Personal Communications*, 115(1):859–876, 2020.

[40] Pukar Maharjan, Kumar Shrestha, Trilochan Bhatta, Hyunok Cho, Chani Park, Md Salauddin, M Toyabur Rahman, SM Sohel Rana, Sanghyun Lee, and Jae Y Park. Keystroke dynamics based hybrid nanogenerators for biometric authentication and identification using artificial intelligence. *Advanced Science*, page 2100711, 2021.

[41] Neenu Daniel and A Anitha. Texture and quality analysis for face spoofing detection. *Computers & Electrical Engineering*, 94:107293, 2021.

[42] Xiaobai Li, Jukka Komulainen, Guoying Zhao, Pong-Chi Yuen, and Matti Pietikäinen. Generalized face anti-spoofing by detecting pulse

from face videos. In *2016 23rd International Conference on Pattern Recognition (ICPR)*, pages 4244–4249. IEEE, 2016.

[43] Si-Qi Liu, Xiangyuan Lan, and Pong C Yuen. Remote photoplethysmography correspondence feature for 3d mask face presentation attack detection. In *Proceedings of the European Conference on Computer Vision (ECCV)*, pages 558–573, 2018.

[44] Zinelabidine Boulkenafet, Jukka Komulainen, and Abdenour Hadid. Face anti-spoofing based on color texture analysis. In *2015 IEEE international conference on image processing (ICIP)*, pages 2636–2640. IEEE, 2015.

[45] Zinelabidine Boulkenafet, Jukka Komulainen, and Abdenour Hadid. Face spoofing detection using colour texture analysis. *IEEE Transactions on Information Forensics and Security*, 11(8):1818–1830, 2016.

[46] Le-Bing Zhang, Fei Peng, Le Qin, and Min Long. Face spoofing detection based on color texture markov feature and support vector machine recursive feature elimination. *Journal of Visual Communication and Image Representation*, 51:56–69, 2018.

[47] Xiaoguang Tu, Zheng Ma, Jian Zhao, Guodong Du, Mei Xie, and Jiashi Feng. Learning generalizable and identity-discriminative representations for face anti-spoofing. *ACM Transactions on Intelligent Systems and Technology (TIST)*, 11(5):1–19, 2020.

[48] Xiao Yang, Wenhan Luo, Linchao Bao, Yuan Gao, Dihong Gong, Shibao Zheng, Zhifeng Li, and Wei Liu. Face anti-spoofing: Model matters, so does data. In *Proceedings of the IEEE/CVF Conference on Computer Vision and Pattern Recognition*, pages 3507–3516, 2019.

[49] Yaojie Liu, Joel Stehouwer, Amin Jourabloo, and Xiaoming Liu. Deep tree learning for zero-shot face anti-spoofing. In *Proceedings of the IEEE/CVF Conference on Computer Vision and Pattern Recognition*, pages 4680–4689, 2019.

[50] Rui Shao, Xiangyuan Lan, Jiawei Li, and Pong C Yuen. Multi-adversarial discriminative deep domain generalization for face presentation attack detection. In *Proceedings of the IEEE/CVF Conference on Computer Vision and Pattern Recognition*, pages 10023–10031, 2019.

[51] Daniel Pérez-Cabo, David Jiménez-Cabello, Artur Costa-Pazo, and Roberto J López-Sastre. Deep anomaly detection for generalized face anti-spoofing. In *Proceedings of the IEEE/CVF Conference on Computer Vision and Pattern Recognition Workshops*, pages 0–0, 2019.

[52] Ian R Fasel, Marian Stewart Bartlett, and Javier R Movellan. A comparison of gabor filter methods for automatic detection of facial landmarks. In *Proceedings of Fifth IEEE international conference on automatic face gesture recognition*, pages 242–246. IEEE, 2002.

[53] Xiangxin Zhu and Deva Ramanan. Face detection, pose estimation, and landmark localization in the wild. In *2012 IEEE conference on computer vision and pattern recognition*, pages 2879–2886. IEEE, 2012.

[54] Manik Varma and Andrew Zisserman. A statistical approach to texture classification from single images. *International journal of computer vision*, 62(1-2):61–81, 2005.

[55] Tony Lindeberg. Scale invariant feature transform. 2012.

[56] Keyurkumar Patel, Hu Han, and Anil K Jain. Secure face unlock: Spoof detection on smartphones. *IEEE transactions on information forensics and security*, 11(10):2268–2283, 2016.

[57] Murat Olgun, Ahmet Okan Onarcan, Kemal Özkan, Şahin Işik, Okan Sezer, Kurtuluş Özgişi, Nazife Gözde Ayter, Zekiye Budak Başçiftçi, Murat Ardiç, and Onur Koyuncu. Wheat grain classification by using dense sift features with svm classifier. *Computers and Electronics in Agriculture*, 122:185–190, 2016.

[58] Engin Tola, Vincent Lepetit, and Pascal Fua. Daisy: An efficient dense descriptor applied to wide-baseline stereo. *IEEE transactions on pattern analysis and machine intelligence*, 32(5):815–830, 2009.

[59] Mahesh Jangid and Sumit Srivastava. Gradient local auto-correlation for handwritten devanagari character recognition. In *2014 International Conference on High Performance Computing and Applications (ICHPCA)*, pages 1–5. IEEE, 2014.

[60] Navneet Dalal and Bill Triggs. Histograms of oriented gradients for human detection. In *2005 IEEE computer society conference on computer vision and pattern recognition (CVPR'05)*, volume 1, pages 886–893. Ieee, 2005.

[61] Jukka Komulainen, Abdenour Hadid, and Matti Pietikäinen. Context based face anti-spoofing. In *2013 IEEE Sixth International Conference on Biometrics: Theory, Applications and Systems (BTAS)*, pages 1–8. IEEE, 2013.

[62] Herbert Bay, Andreas Ess, Tinne Tuytelaars, and Luc Van Gool. Speeded-up robust features (surf). *Computer vision and image understanding*, 110(3):346–359, 2008.

[63] Yaman Akbulut, Abdulkadir Şengür, Ümit Budak, and Sami Ekici. Deep learning based face liveness detection in videos. In *2017 international*

artificial intelligence and data processing symposium (IDAP), pages 1–4. IEEE, 2017.

[64] Abdulkadir Şengür, Zahid Akhtar, Yaman Akbulut, Sami Ekici, and Ümit Budak. Deep feature extraction for face liveness detection. In *2018 International Conference on Artificial Intelligence and Data Processing (IDAP)*, pages 1–4. IEEE, 2018.

[65] Aziz Alotaibi and Ausif Mahmood. Deep face liveness detection based on nonlinear diffusion using convolution neural network. *Signal, Image and Video Processing*, 11(4):713–720, 2017.

7

Enhanced Tal Hassner and Gil Levi Approach for Prediction of Age and Gender with Mask and Mask less

Srikanth Busa[1], Jayaprada Somala[2], T. Santhi Sri[3], and Padmaja Grandhe[4]

[1]Department of Computer Science & Engineering, Kallam Haranadhareddy Institute of Technology, India
[2]Department of Computer Science & Engineering, Lakireddy Bali Reddy College of Engineering, India
[3]Department of Computer Science & Engineering, Koneru Lakshmaiah Education Foundation, India
[4]Department of Computer Science & Engineering, Potti Sriramulu Chalavadhi Mallikarjunarao College of Engineering & Technology, India
E-mail: srikanth.busa@gmail.com; jayasomala@gmail.com; santhisri@kluniversity.in; padmajagrandhe@gmail.com

Abstract

During this pandemic scenario, most people are getting vaccinated by violating the age rules defined by the. We found a few scenarios where persons aged 30–40 have claimed their age as 45 and got vaccinated during the early days of vaccination. In some scenarios, students aged 18–20 also got vaccinated before the government declared vaccination for those above 18 years. By considering all these scenarios, the proposed system has implemented a model that can recognize the person's gender and age for verification before allowing them to register for vaccination. Previously, different models could recognize the face along with their age and gender. However, this system proved its efficiency by predicting accurate results with and without masks. This system has achieved an accuracy of "98.51%," which is far better than the previous models stated in the literature survey. The proposed model has utilized the pretrained model best for age and gender are known as "Tal Hassner and Gil Levi." The existing THGL CNN model consists of five-layered architecture in which three layers are CNN layers, and two layers are dense. In the existing model, first two layers take care of the normalization process with the help of local search algorithms, and the third layer implements 384 filters to extract the necessary features. The first dense layer uses these extracted features and uses 512 neurons to predict the gender of the person using ReLu activation. The last layer uses the softmax layer because age is a multi-classification process. This model tested the application using both cropped images and over-sampled images, and the average prediction rate is taken into account.

Keywords: Tal Hassner and Gil Levi, Caffe model, network configuration, prediction, HAAR, pooling, normalization, Swish.

7.1 Introduction

The proposed chapter aims to identify the age and gender of the person at the COVID vaccination centers to help the government recognize the people who want to get vaccinated at the wrong age. Identifying age and gender is a popular application in many social media applications, but the wrong prediction of age will not impact people. In a pandemic situation, it is crucial to identify the person's age for two reasons: one is due to scarcity of the vaccine at the early stages of the situation, and another is to decide on the right people to get vaccinated. So to handle these situations, the proposed model gets motivated to predict the correct age and gender of the person. The proposed method uses 11-layered architecture [1], in which it uses five CNN layers, three max pooling layers, and three fully connected layers. In the past few years, CNN [2] has been the most audible term coined by most researchers, industry people, and other prominent persons of AI. CNN is a deep learning approach that is used to differentiate among the various objects of inputs based on the prioritized weights [3] assigned to them. The main advantage of these neural networks lies in their recognition process of dependencies among the spatial and temporal features using different kinds of filters. The major goal of any network is to understand the input images with less trainable parameters. In a real-time scenario, we can see images of different color systems like RGB, CMYK, and others [4], so there is a need for a network that can efficiently convert the image into an easy processing

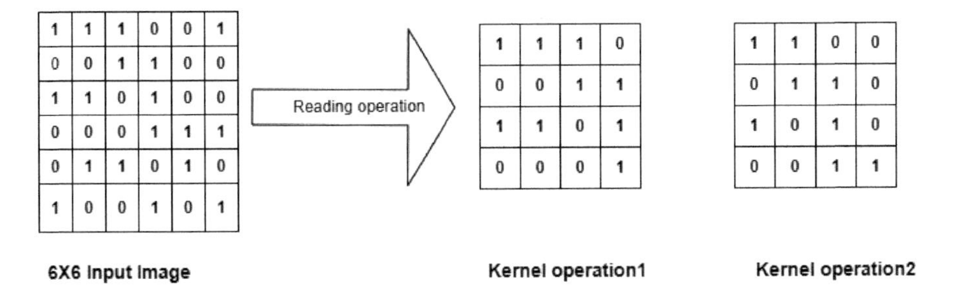

Figure 7.1 Reading operation by kernel component.

format. However, still, it needs to maintain all its important features to get accurate prediction results. The major components of networks are:

1. Kernel: This also known as filter [5], which is used to read the pixel values of the image in stage by stage manner. The number of pixels, it needs to process is dependent on the size of the kernel. Suppose we have K as $4\times4\times1$ then, the reading of complete image needs 16 shift operations by reading four values horizontally and vertically, this can be pictorially represented as shown in Figure 7.1. The main aim of the convolution layers [6] is to obtain the convoluted matrix, which contains information about the high-level features. With the increase in layers, we can obtain low-level features like color, gradients, and others.
2. Padding: It is an attribute that decides the size of the output obtained. If the padding attribute is considered valid, then it returns the output matrix with reduced dimensions [7]. If the padding attribute is considered the same, it returns the output matrix with the same number of dimensions as in the case of input.
3. Pooling layer: The time complexity of the system increases with the increase of features in the image. To preserve all the important features with high-speed processing [7], the CNN makes use of this pooling layer. It also reduces the computational power of the system. The proposed system has implemented a max pooling layer to extract the features from the image.
4. Fully connected layer: Before sending the input to this layer, the system must flatten the matrix to resize the matrix into the necessary dimensions. After receiving the flattened values, the task of this layer is to identify the attributes that have non-linear relations between

them. The proposed system used both feedforward and backpropagation technique along with a softmax activation function to classify the age of the persons [8].

Problems and challenges: while the model is capturing the face of the person at the entrance of the COVID vaccination center it may face following problems:

1. Collection of the real images is difficult to design a good amount dataset because the conventional face detection model uses the faces of foreigners or anglo Americans, based on which it is highly difficult to predict the ages of the Indian people. There are many variations in terms of complexion, hair nature, and eye iris color.
2. During the capturing process, the eyes of the person might get closed due to the flashlight of the system. In general, many traditional systems have used eyes as the major recognition element, i.e., iris recognition is the famous design for facial detection.
3. Due to the low light intensity, blurriness may be there in the face recognition process-the blurriness results in noisy data. The traditional approaches like wiener filter, Gaussian filter, or salt and pepper noise reduction techniques are applied. These traditional approaches are not efficient in making the prediction process efficient.
4. To predict the gender, the conventional approaches use the deployment caffee model. Using the deployment context, it is difficult to predict the person's exact age in terms of years and months.

The proposed research designs its own dataset by integrating the real-world images of the Indians with the dataset available at [9] for training purposes. The proposed research uses CNN instead of manual image processing techniques, famous for automatic feature extraction to solve iris recognition and low-intensity problems. The layers of the CNN are trained in such a way that they can handle any noise and are experts in extracting the low- and high-end details from the image by applying different types of kernels based on image size. The major novelty of the proposed research lies in detecting the facial features using the combination of pretrained models and HAAR. It helps the designed model predict the age group range and correct gender classification rather than predicting the wrong age. HAAR transformations are popular for identifying low-level details like corners, edges, and lines, whereas the designed THGL model, using 11-layered architecture, focuses on the higher-level details. So, the overall proposed model checks every detail with a high accuracy rate.

7.2 Literature Review

This research study [10], a DL technique for systemic and ocular disorders based on pictures from the fundus. The retina has characteristics that can influence gender distinctions and depends on the retinal imaging region in which these characteristics are recorded through the photo. Their objective is to propose a comparative study in DL algorithms' performance to predict the gender depending on the various fields of fundus images. A rigorous cross-examination was performed over various patients of age groups more than and equal to 40. Various images of fundus were collected in this study as an input for the proposed DL framework for predicting age. The performance was calculated at the image and individual levels, and ROC curves were opted to know binary classification. AUC curves were also opted to classify the image characteristics at both image and individual levels. The researcher states that their best performance was seen at optic decentered field photos with AUC of 0.91.

The objectives of this research paper were to identify the degree of learner studies utilizing ML algorithms on the technologies, ICTs, and MTs [11]. Two distinct categories such as UG and PG students, are categorized. We employed main datasets based on four supervised classifications in IBM SPSS Modeler 18.2 in forecasting these two categories. An auto classifier algorithm was also introduced to develop a model, namely, XGBT, RT and RF, and the other ML-supervised algorithms. Apart from these, a SMOTE methodology was also designed to maintain the dataset balanced in advancing the forecasting accuracy. The outcomes of the proposed methodology state that the XGBT methodology has shown the highest performance of 92.36% over the 16 significant properties at the stud stage prediction. On the one hand, there has been a substantial divergence between XGBT and RT reliability, and on the other hand, RF and XGBT are not affected significantly.

According to Verma *et al.* [12], GBM is said to be the most popular and dreadful brain tumor specifically seen in adults. A rigorous study has been conducted on the GBM patients developing a DL framework using the techniques RSF and CoxPH. Including this, a study was conducted to observe the effects of the hyperparameters on the optimization methods. The model to improve the foresight efficiency of the DL-based survival methodologies. The experimental study was conducted on patients with GBM by considering various factors and their effects, and the machine's performance was calculated accordingly to those factors. CoxPH is the most common survival technique in medical studies as its execution is simple in an informative explanation. Another methodology named DeepSurv, a DL technique that is also an extension to the CoxPH model, was also introduced to deal with the risk factors in the model. The study concludes by stating that the GBM is lesser in the age groups less than 40 by demonstrating the CoxPH model's performance that derives 95% with the affecting factor age.

According to the Moradmand *et al.* [13], some organizations need their customer's ages and gender in order to provide more enhanced services to meet the client's requirements. Thus, they depend on their enterprise's system to gather the users' information. This data needs to be precise, and, in a few cases, it must be in the desired format. Therefore, some regulations were taken up from the validators, and this works well with various types of data. There is an issue regarding the ignorance of detecting the user's age and gender. They also do not foresee the users' age or gender from their IDs. The proposed research is determined to identify the user's age and gender based on their ID, i.e., from the photo attached to that ID precisely. A CNN or ConvNets approach with a double-verification layer that associates with the user's photo, gender, and the birth period from the given ID proofs, to estimate the user's age and gender was developed. Later, a service through the web was designed to make a verification process. Finally, the designed approach showed 82.2% accuracy in gender prediction and 57% in age prediction.

From the view Abu Nada *et al.* [14], there is a massive urge with engaging online platforms, and the tremendous, short texts that we are generating and are circulating over the Internet create an obligation to recognize the age and sex of that individual automatically who circulates the messages. Various websites try to misguide and deceive the individuals who visit them by contributing fraudulent info about their gender and age. Many standard methodologies in detecting were based on manual works which lack perfection and are no longer useful. Thus, this research paper proposes an automatic tool with unique properties to analyze the selected phrase. The properties include many characteristics based on linguistics. The author states that their designed framework with some production rules had outperformed the standard detecting measures. With the features CFG, the proposed SVM classifier has achieved a precision of 82.81% with gender and 83.2% of perfection in maturity predictions with the same SVM classifier.

According to the Xin *et al.* [16], predicting mind age based on visual input and ML algorithms has significant possibilities to deliver knowledge into cognitive and mental illness progression. While various ML approaches have been developed, it is still required to systematically compare ML techniques with imagery attributes obtained from diverse modalities. In this work, 36 permutations of imagery characters and ML models incorporating DL are assessed for prediction effectiveness. The author claims that the standard approaches have shown a methodical variance in predicting the mind's maturity level. The non-sequential relationship between the mind and the periodical age was not tested intensely. Here, they had suggested a new way to address the consistent variation of brain age gaps by considering factors of gender, biological age, and their interconnections. Because the actual brain age is unpredictable and may depart from actual age, they subsequently investigate if different degrees of behaviors, based upon approximated neuroimaging information, anticipate their brain age. SVR, GPR, DNN, and RR were implemented with the performance metrics R2 and MAE.

Singh *et al.* [17] developed a method bases on automatic age plus gender identification. It has been widely used in everyday lives, most notably in human-computer interaction, visual footage, fingerprint evaluation, computers, and other business applications. The author can develop the recommended systems by recognizing people's feelings that have already existed. These models perform admirably on actual pictures. When the facial expressions of the source images are unbiased or quiet, it performs poorly for age prediction. For age and gender model classification, the data is gained at the dataset IMDb-WIKI, and as well as for stress detection, the Fer2013 data has been obtained from Kaggle. The above application has two concepts, age and another one is gender prediction utilizing a broad resent architectural style, and expression recognition utilizing the traditional CNN model. The use of CNN resulted in the development and improvement of these activities. The present method achieves an accuracy of about 96.26%, while the sentiment analysis model achieves an accuracy of about 69.2%.

Bremberg *et al.* [18] implemented a model using CNN. This study aims to categorize the age and gender using the face in the picture is showing. The purpose is to show off a method specializing in characterizing historical photos and functionality where consumers could put images for analysis. This chapter will make studying history easier by developing new picture analysis tools. The methodology was written in Python by using the convolutional neural networks to calculate the age and gender of the people. The user interface is built with JavaScript and React to communicate with algorithms. The gender identification and classification algorithm do have a 96% accuracy rate, and as well as age identification algorithm has an average age mistake of 4.3 years. Our study also revealed that the method outperformed the frequently utilized art classifier method on history images.

Benkaddour *et al.* [19] designed a model for pattern identification and automatic classification. These seem to be more active research zones, with the main goal of developing intelligent machines capable of learning and recognizing items effectively. Biometric technologies can be used for security reasons. Overall, it is an important component of these apps. The face method has grown in importance as a foundational biometric technique in this area of investigation. Using convolutional neural networks, this analysis aims to create a gender forecasting and age evaluation framework for facial images or actual videos. Inside this article, three convolutional neural methods with various architectures have been created and validated on IMDB and WIKI datasets. The analysis indicates that CNN channels have greatly improved their performance with accuracy.

Virmani did some research and developed a CNN model. Identifying human feelings through facial expressions is known as identifying facial expressions. The idea has grown in popularity and researchers have been trying to broaden the scope by developing suitable ways for extracting more information from face characteristics. As a result, this chapter introduces numerous facial recognition systems that can analyze human face features to foresee human feeling, phase, gender, and age. Towards the full knowledge, not only one paper describes a system that can detect all these characteristics simultaneously in a single image. The foundation of the complete project is composed of Haar feedback and a convolution neural network. Assistance for a Web app utilizing flask software has also been provided, allowing the task to be implemented on some other desktop using the same channel through the browser. This results in providing an accuracy of 68.33%.

Guo *et al.* [4] have proposed a model to estimate the age. It is a significant challenging issue in various areas, such as migrant's identifier, relevant laws, and medical treatments. The deep learning model has recently been used to identify age and prediction. However, there is no comparable performance among both manuals and ML models. It depends on a large survey of dentists' orthopantomograms (OPGs). For such research, the author collected a total of 10,257 orthopantomograms. To compare the manual process, the author created logistic and linear methods for each age at the final convolution neural network (CNN) that categorized the teeth age immediately. These processes depend on the left mandible eight permanent dentition or the third tooth individually. Our findings show that when especially in comparison to manual processes (92.5%, 91.3%, and 91.8% for age), finally CNN methods perform better accuracy for 14, 16, as well as 18 age people (95.9%, 95.4%, 92.3% for the age of 14, 16, as well as 18) [3].

Kuppusamy developed the CNN model. With the introduction of conversational vocal identification systems such as Amazon Echo, SIRI, Google, and others, general language interactive schemes like Chabot and voice control by identification of the systems are at an all-time high, and defining the age of either a person speaking is crucial for establishing the relevant context. Age could be deduced from an audio signal by using numerous factors, including physical speech characteristics, linguistics, regularity, and voice quality This study describes using deep learning methods to retrieve spectral characteristics of speaking as input data, including campestral coefficients, spectroscopy reduce, center of gravity, Uniformity,and spatial frequency. A new approach and the design for an application are discussed for categorizing the characteristics of utilizing deep neural networks and convolutional neural networks. By utilizing these three methods, the proposed methodology results can get a great performance rate.

Table 7.1 represents the merits and limitations that are caused by the previously proposed systems.

7.3 Proposed Methodology

The proposed system has worked on its dataset, which is designed as an integration of public dataset available [9] and real-world images. Instead of traditional image processing techniques [20], the proposed system wants to detect maturity and gender with the use of a pretrained CNN model known as "Tal Hassner and Gil Levi" The complete architecture of the proposed system is shown in Figure 7.2. The system initially performs the preprocessing techniques like removing noisy data and resizing all the images to exact size. The model increases the size of the dataset by performing data augmentation operations on the images like rotation, transformation, scaling, and others. These newly created images, collected ones, and the publicly available dataset images are passed as input to the HAAR model. The major goal of the HAAR model [21–24] is to detect the face by subtracting the noise and background portion. The entire process is divided into three parts, as discussed below.

1. Face recognition using HAAR cascade: In a 2D image or 3D range image, facial features extraction is the process of locating certain regions, points, landmarks, or curves/contours. In this feature extraction

Table 7.1 Merits and demerits of previously developed models.

S.No.	Author	Algorithms used	Merits	Demerits and future work
1.	Betzler	CNN .	Concentrated on deriving the most impacting criteria for gender prediction.	In future, gender related diseases can also be derived using external verification sets.
2.	Verma	XGBT, RT, RF and SMOTE.	A stage vice experimental analysis was performed with variant features.	NN models over the experiment may show greater results with a deeper analysis.
3.	Moradmand	DeepSurv, DeepSurv Bayesian, RSF, CoxPH	Surviving methodologies in DL were opted for deriving the optimal results.	Further study includes dealing with the reproducible radiomics properties for the DL survival model.
4.	Nada	CNN, OpenCV	Gender prediction evaluation is greater. Age prediction lacks.	More attributes and data can be considered to make the system more compatible.
5.	Abdallah	SVM,NB and DT	A gain feature selection was implemented to achieve greater accuracy and to remove unnecessary features.	DL techniques or combinatorial algorithms may retrieve higher results.
6.	Niu	RR, SVR, DNN and GPR.	The usage of multi-modal features has shown best results.	In future, the model can focus on maturing and the effects of the brain's aging can be considered.
7	Singh	CNN and Haar cascades classifier.	Classification has done based on expression.	Entity reorganization has not done and more layering network.
8	Bremberg	Histogram of oriented gradients, CNN, pattern recognition.	Feature extraction and image featuring can be determined.	To enhance the recognize rate by noise, background.

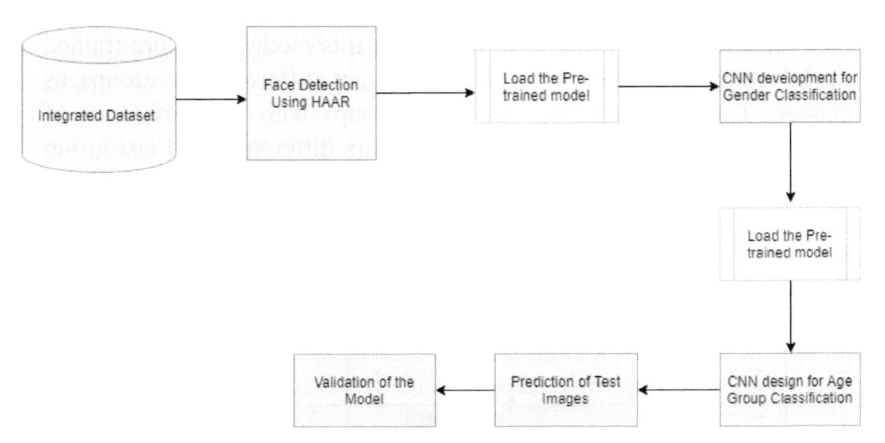

Figure 7.2 Architecture for enhanced THGL.

stage, the registered image is used to generate a numerical feature vector. Some of the most common qualities that can be extracted are as follows: Mouth, cheekbones, eyelids, eye, and nose suggestion. HAAR is comparable to kernels, which is a feature that is frequently used to identify edges. All people's facades have the same looks: the areas around and at the nose and the upper cheeks of a person would be usually brighter than the region around their eyes. We will be able to detect a face based on the size and placement of these matchable features. The components shown are used to detect edge and line characteristics as in Figure 7.3. Here are a couple of feature vectors to check for the presence of a face. In the HAAR feature, +1 represents the dark region while –1 represents the light zone. In a 24×24 window, a picture is displayed. To produce each feature, the total number of pixels in the bright and dark boxes is subtracted from the total number of pixels in the white and black rectangles. Many features are now calculated with all possible kernel sizes and placements. The system must calculate the entire count of pixels under the bright and dark boxes for each feature computation.

2. Gender recognition with neural networks: Here, the implemented model is known as "Caffe-m-porter-NET," which contains the following network architecture with softmax as activation function in the second fully connected layer which classifies the person as either male or female and the process is illustrated in Figure 7.4.

3. Age recognition with neural networks: In this module, the pre-trained model contains multi classification of eight different age groups as classes. Those eight groups are: two groups with age difference of two years, five groups with a range of years differently and last group contains 60–100 years age people. To predict the age [25, 26] group of the persons, it considers the same architecture as of gender prediction but in the last year instead of softmax activation function, it used forward function since it is a multi classification process.

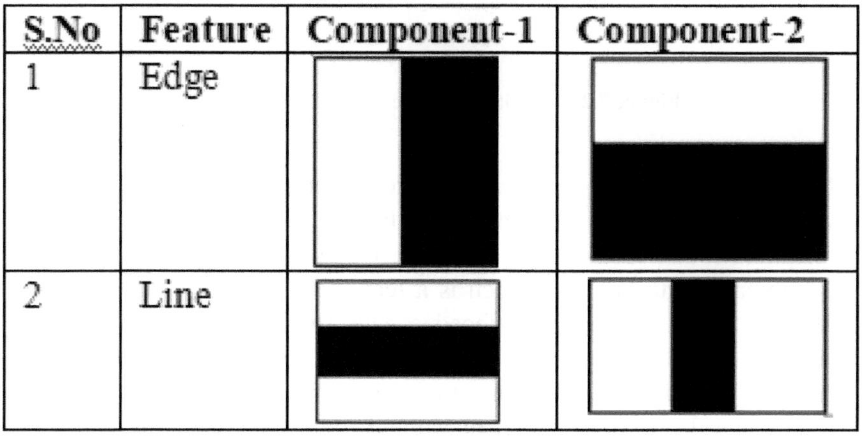

S.No	Feature	Component-1	Component-2
1	Edge		
2	Line		

Figure 7.3 Components of HAAR.

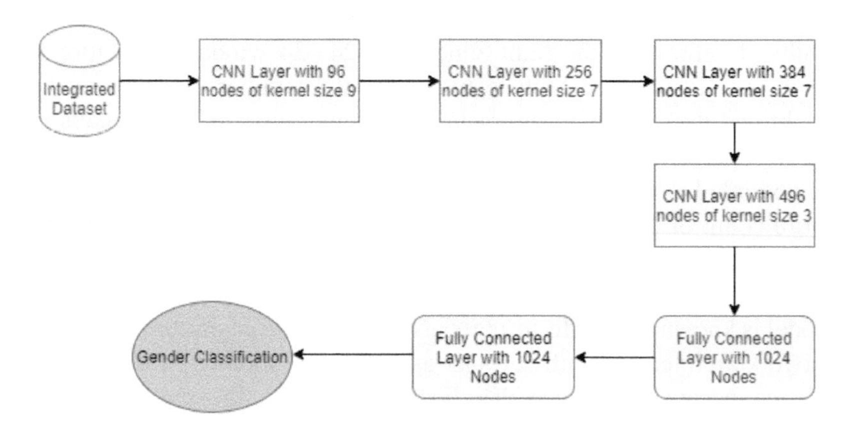

Figure 7.4 Network architecture for gender classification.

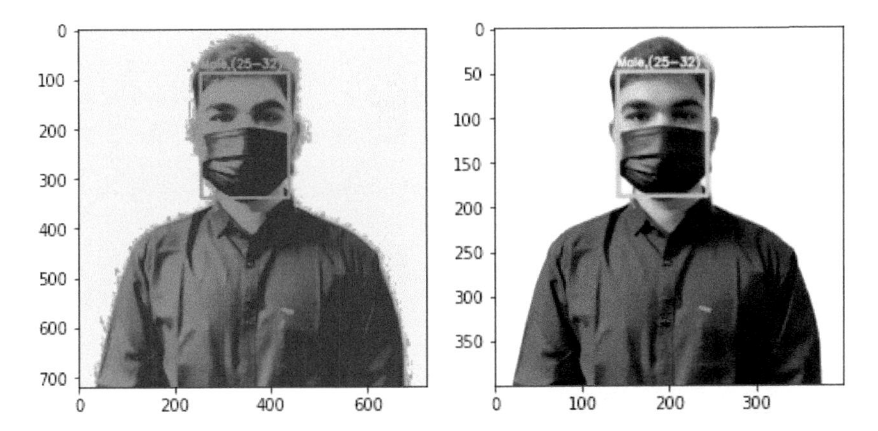

Figure 7.5 Presentation of output for histogram equalization and resize operations.

7.3.1 Caffe model

It is a popular neural network model to build a classifier in simple three steps.

1. Preprocessing: This method is executed using the cloud services of Amazon Web Services. During this phase, it applies histogram equalization operation on all the training images to enhance the quality of the pictographs, all the images need to be resized to 350×300 size. The output of the preprocessing operations was presented in Figure 7.5 The equalization process is achieved using the limited contrast technique (CLAHE) in the proposed system. Because it can efficiently handle the noise amplifications caused by various factors, it stores the data in two types of databases with the extension ".lmdb," one for training and one for validation. It is a lightweight memory-mapped database, making the execution rate as fast as possible in AWS as executing on the localhost. The output of this preprocessing step is stored as a "datum" file, which returns a label associated with the image.
2. Model configuration using NN: The main aim of this step is to define the necessary parameters required by each layer. It uses AlexNet, which is modified as shown in Figure 7.6. All the images are resized to 128×128 and passed through the first CNN layer. The significant advantage of CNN is to design the extracted features with low dimensionality without loss of information. The initial size of the image is 128 × 128 × 3 = 49,152. The first layer is an input layer to construct a single-dimension

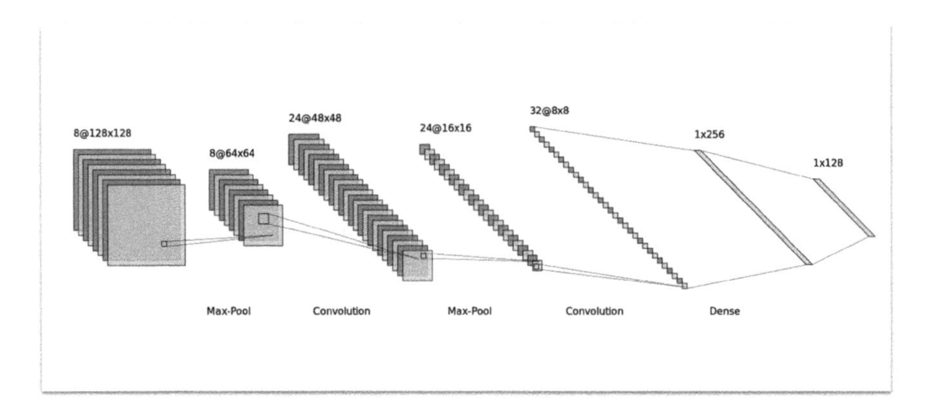

Figure 7.6 Neural network model configuration.

matrix based on the image size. Here, since the images are resized 128 × 128 = 16,384, the system has 1000 samples. Finally, the dimension of the first layer is (16,384, 1000). In the proposed research, the second is designed to detect the edges in horizontal and vertical directions. To perform edge detection, it applies the Eq. (7.1)

$$(IS * IS) * (FS * FS) = (IS - FS + 1)^2. \tag{7.1}$$

In Eq. (7.1), IS indicates image size, and FS indicates filter size. The pixel values of the CNN layer data are filled with dot value computations. This layer applies an activation function called "ReLu" to normalize the negative values to zero. The common and in-between layers of CNN are pooling layers whose goal is to reduce the dimensionality in terms of volume. By changing the stride and filter size, the volume of the image changes dynamically. Out of the three last layers, two layers, i.e., dense layers, are used for classification purposes by adjusting the weights and biases and defining the number of neurons. The last layer used is the softmax layer because the age group is not a binary value.

3. Solver: The solver task is to determine the model's accuracy by validating it against the different cross-folded parts of the dataset. The optimization values of this process are based on the parameters like learning rate, epochs, and gamma values. The learning rate or step size computes the rate of updation in terms of weights after each epoch completion. The best values for this parameter always lie between 0 and

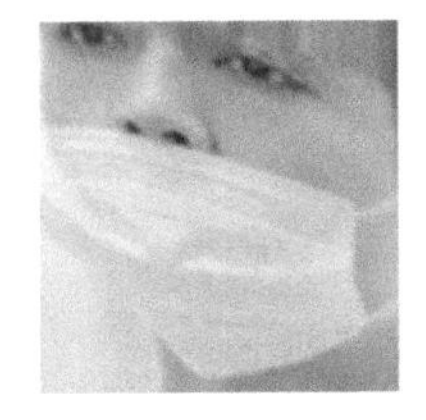

Existing Dataset in the Kaggle

Self Collected Images from the real world.

Figure 7.7 Dataset proposed in the research.

1. The learning rate should always be an average case because low values make the model learn features with more training, whereas high values make the model quickly learn the weights by converging at some point in the network. One complete iteration with both forward and backward processes is known as an "epoch," used to train the model. The parameter epoch defines the number of times a model has to undergo the training process to learn all the data attributes.

7.4 Experimental Results

7.4.1 Materials used

In the proposed research, the model created its dataset by scraping the data from the real-world images and it is integrated with the dataset [9] available in the Kaggle. Figure 7.7 describes and represents the dataset utilized in the research. This data repository contains three types of persons one with a mask, one without a mask, and others not properly wearing the masks. The dataset is a balanced dataset; during the integration process, equal distribution of images to each class was also added. Finally, regarding the facade identification, the system used a .pb file representing a protocol buffer that usually maintains the grid purpose and the weights that the model trained. One .pb file carries the data in binary type, and the other .pbtxt holds the same data in text format. These files are generally derived from TensorFlow for

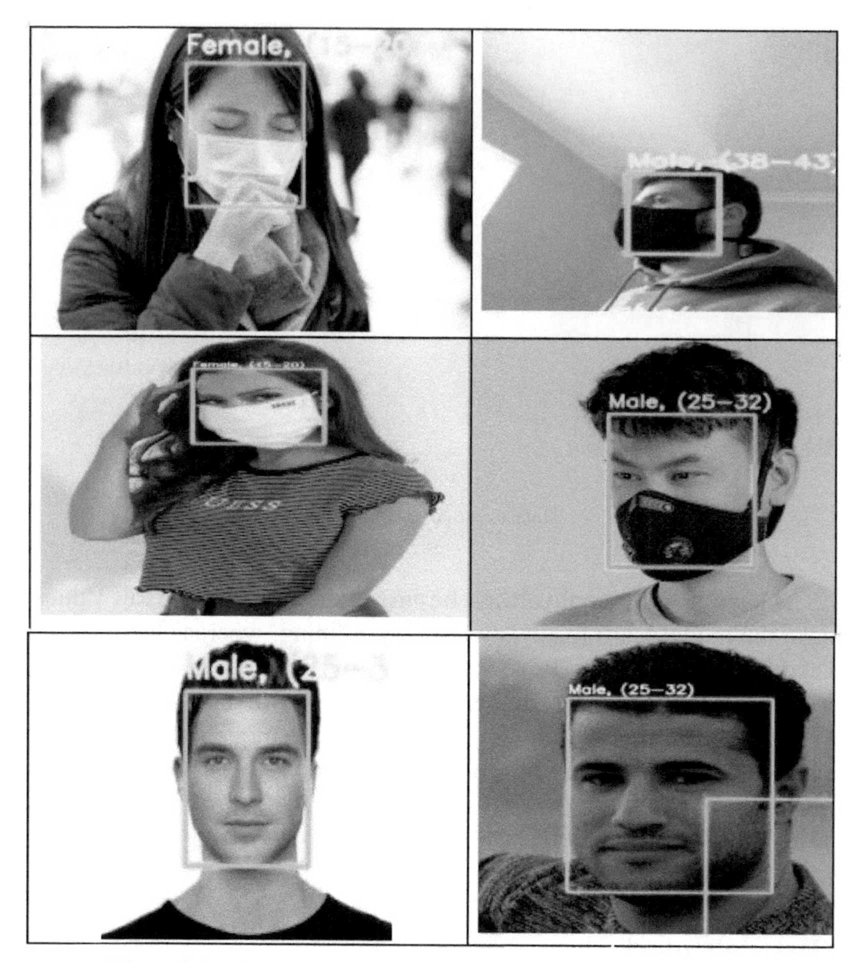

Figure 7.8 Age and gender prediction with mask and without mask.

determining maturity and gender. A .prototxt file is maintained that regulates the network information and whereas .caffe type of file manages the internal states of the attributes of the layers.

Figure 7.8 determines the age and gender of the persons with mask and without mask.

7.4.2 Results and discussion

The proposed research is implemented in the Google Colab environment because the dataset contains many images. The system needs a high-end

Table 7.2 Estimators assumptions for CNN.

SI.No.	Estimator name	Estimator value
1	Number of layers	11
2	Number of neurons in the hidden layer	128
3	Batch size	35
4	Learning rate	0.08
5	Activation function	ReLu
6	Optimizer	Adam

processor to use GPUs available in the collaboratory to work with these images. To implement the proposed algorithm, it needs TensorFlow and KERA's library. All the images are stored in Google drive. It is accessed with the help of a session token generated by the Colab. Table 7.2 represents the assumed parameters for designing the neural network.

By executing the proposed algorithm, and setting the epochs to 150, the training takes as shown in Figure 7.9. Figure 7.9 represents the amount of loss and accuracy obtained by the training images after each epoch. The total training set images are divided into various batches, each of size 35. In each iteration, it observes time taken to train the images, loss incurred, and accuracy. From the Figure 7.9, it is clear evident that with the increase of

```
Epoch 1/150
35/35 [==============================] - 1s 3ms/step - loss: 218.7020 - accuracy: 0.0690
Epoch 2/150
35/35 [==============================] - 0s 4ms/step - loss: 19.8357 - accuracy: 0.4713
Epoch 3/150
35/35 [==============================] - 0s 3ms/step - loss: 4.6094 - accuracy: 0.7155
Epoch 4/150
35/35 [==============================] - 0s 3ms/step - loss: 4.0239 - accuracy: 0.6897
Epoch 5/150
35/35 [==============================] - 0s 3ms/step - loss: 3.3229 - accuracy: 0.6925
Epoch 6/150
35/35 [==============================] - 0s 2ms/step - loss: 2.6924 - accuracy: 0.6954
Epoch 7/150
35/35 [==============================] - 0s 3ms/step - loss: 2.0524 - accuracy: 0.7213
Epoch 8/150
35/35 [==============================] - 0s 3ms/step - loss: 1.4776 - accuracy: 0.7270
Epoch 9/150
35/35 [==============================] - 0s 3ms/step - loss: 0.9347 - accuracy: 0.7385
Epoch 10/150
35/35 [==============================] - 0s 3ms/step - loss: 0.5260 - accuracy: 0.7586
Epoch 11/150
35/35 [==============================] - 0s 3ms/step - loss: 0.3455 - accuracy: 0.7672
Epoch 12/150
35/35 [==============================] - 0s 3ms/step - loss: 0.3245 - accuracy: 0.7672
Epoch 13/150
35/35 [==============================] - 0s 3ms/step - loss: 0.2780 - accuracy: 0.7730
```

Figure 7.9 Training process to classify the images.

Table 7.3 Estimators assumptions for CNN.

S.No	Algorithms used	Accuracy	Recall	Precision	F1-Score
1	CNN	91	89.9	88.1	89.25
2	XGBT, RT, RF and SMOTE	92.36	95.6	94	94.2
3	DeepSurv, DeepSurv Bayesian	95	95	95	95
4	CNN, OpenCV	82.2	79.3	80.14	79.8
5	SVM,NB and DT	83.2	80.4	81.42	80.7
6	RR, SVR, DNN and GPR	89.1	88	87.5	87.5
7	Enhanced THGL	98.51	96.7	97.2	97.05

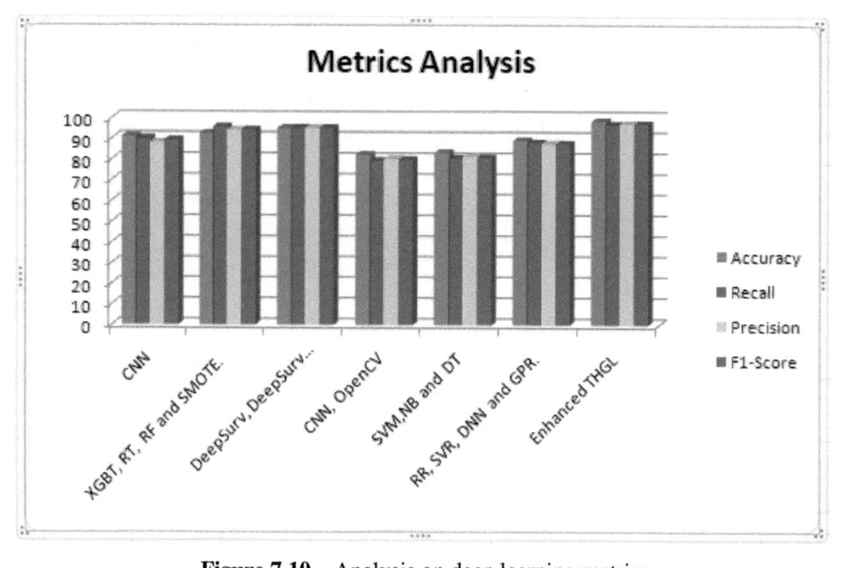

Figure 7.10 Analysis on deep learning metrics

epochs, the loss rate reduces and accuracy rate increases and at the same time train the images also get reduced. Table 7.3 represents the various parameters associated proposed algorithm and is compared with the various existing algorithms. Figure 7.10 represents comparison graph of existing approaches with the proposed algorithm by representing algorithms on X-axis and measurements on the Y-axis. It is clear that in terms of all metrics, the proposed algorithm performs better.

7.5 Conclusion

The proposed system can efficiently detect the era and sex of the person with and without a mask and helps the vaccination centers. To take the necessary steps to stop the misutilization of the vaccines and make sure that it reaches the people based on their correct age factors only. Tal Hassner and Gil Levi's enhanced model modifies a few layers, number of neurons, and activation function to attain an accuracy of 98.51%, which is better than the existing one, i.e 84.7%. In future work, the system can be designed to train the images of transgender and try to predict them along with their age. The system can also try to predict the correct age rather than print the age range. Few boundaries need attention like a strict age display like 44 years 5 months will help further research avoid an even small amount of vaccine for correct usage.

References

[1] James Rwigema, Joseph Mfitumukiza, and Kim Tae-Yong. A hybrid approach of neural networks for age and gender classification through decision fusion, Apr 2021.

[2] Speaker age and gender estimation based on deep learning bidirectional long-short term memory (bilstm). *Tikrit Journal of Pure Science*, 26(4):76–84, 2021.

[3] Karthika Kuppusamy and Chandra Eswaran. Convolutional and deep neural networks based techniques for extracting the age-relevant features of the speaker, Apr 2021.

[4] Yu-cheng Guo, Mengqi Han, Yuting Chi, Hong Long, Dong Zhang, Jing Yang, Yang Yang, Teng Chen, and Shaoyi Du. Accurate age classification using manual method and deep convolutional neural network based on orthopantomogram images, Mar 2021.

[5] Eren KARAKAŞ and İbrahim YUcel OZBEK. Age group and gender classification from facial images based on deep neural network fusion, Mar 2021.

[6] Tejas Agarwal, Mira Andhale, Anand Khule, and Rushikesh Borse. Age and gender classification based on deep learning. *Techno-Societal 2020*, page 425–437, 2021.

[7] Deepali Virmani, Tanu Sharma, and Muskan Garg. Gaper: Gender, age, pose and emotion recognition using deep neural networks. *Lecture Notes in Mechanical Engineering*, page 287–297, Sep 2020.

[8] Aryan Saxena, Prabhangad Singh, and Shailendra Narayan Singh. Gender and age detection using deep learning. *2021 11th International Conference on Cloud Computing, Data Science & Engineering (Confluence)*, Jan 2021.

[9] S. Vasavi, P. Vineela, and S. Venkat Raman. Age detection in a surveillance video using deep learning technique. *SN Computer Science*, 2(4), Apr 2021.

[10] Vijay Kumar. Face mask detection, May 2021.

[11] Bjorn Betzler, Henrik Hee, Sahil Thakur, Marco Yu, Ten Cheer Quek, Zhi Da Soh, Geunyoung Lee, Yih-Chung Tham, Tien Yin Wong, Tyler Rim, and Ching-yu Cheng. Gender prediction via deep learning across different retinal fundus photograph fields: a multi-ethnic study (preprint). *JMIR Medical Informatics*, 9, 10 2020.

[12] Chaman Verma, Zoltán Illés, and Veronika Stoffová. Study level prediction of indian and hungarian students towards ict and mobile technology for the real-time. In *2020 International Conference on Computation, Automation and Knowledge Management (ICCAKM)*, pages 215–219, 2020.

[13] Hajar Moradmand, Seyed Aghamiri, Reza Ghaderi, and Hamid Emami. The role of deep learning-based survival model in improving survival prediction of patients with glioblastoma. *Cancer Medicine*, 10, 08 2021.

[14] Abdullah M Abu Nada, Eman Alajrami, Ahmed A Al-Saqqa, and Samy S Abu-Naser. Age and gender prediction and validation through single user images using cnn. 2020.

[15] Emad Abdallah, Jamil Alzghoul, and Muath Alzghool. Age and gender prediction in open domain text. *Procedia Computer Science*, 170:563–570, 01 2020.

[16] Niu Xin, Fengqing Zhang, John Kounios, and Hualou Liang. Improved prediction of brain age using multimodal neuroimaging data. *Human Brain Mapping*, 41, 12 2019.

[17] Arjun Singh, Nishant Rai, Prateek Sharma, Preeti Nagrath, and Rachna Jain. Age, gender prediction and emotion recognition using convolutional neural network. *SSRN Electronic Journal*, 01 2021.

[18] Ulrika Bremberg, Liv Cederin, Gabriel Lindgren, and Filip Pagliaro. Classifying age and gender on historical photographs using convolutional neural networks, 2021.

[19] Mohammed Kamel Benkaddour, Sara Lahlali, and Maroua Trabelsi. Human age and gender classification using convolutional neural network, Feb 2021.

[20] Padmaja Grandhe. A novel method for content based 3d medical image retrieval system using dual tree m-band wavelets transform and multiclass support vector machine. *Journal of Advanced Research in Dynamical and Control Systems*, 12:279–286, 03 2020.

[21] Padmaja Grandhe, Sreenivasa Edara, and Vasumathi Devara. Adaptive roi search for 3d visualization of mri medical images. pages 3785–3788, 08 2017.

[22] Padmaja Grandhe, Sreenivas Edara, and Vasumathi Devara. Adaptive analysis & reconstruction of 3d dicom images using enhancement based sbir algorithm over mri. *Biomedical Research (India)*, 29:644–653, 01 2018.

[23] Padmaja Grandhe, Sreenivasa Edara, and Vasumathi Devara. *An Extensive Study of Visual Search Models on Medical Databases*, pages 219–228. 11 2018.

[24] Padmaja Grandhe, Edara Reddy, and D. Vasumathi. An adaptive cluster based image search and retrieve for interactive roi to mri image filtering, segmentation, and registration. 94:230–247, 12 2016.

[25] Saiyed Umer, Bibhas Dhara, and Bhabatosh Chanda. Biometric recognition system for challenging faces. pages 1–4, 12 2015.

[26] Saiyed Umer, Bibhas Dhara, and Bhabatosh Chanda. Face recognition using fusion of feature learning techniques. *Measurement*, 146, 06 2019.

8

A Brief Overview of Recent Techniques in Crowd Counting and Density Estimation

Chiara Pero

Department of Computer Science, University of Salerno, Salerno, Italy
E-mail: cpero@unisa.it

Abstract

Estimating crowd counting from images is a difficult but important task given the large range of applications such as public safety, traffic monitoring, and urban planning. Occlusions, uneven density, variation in scale and perspective are all challenges in crowd analysis. Thanks to advancements in deep learning and constructing demanding databases, modern computer vision techniques have led to numerous cutting-edge methods that build the abilities needed to properly execute a wide range of scenarios. This article presents a brief description of pioneering methods based on hand-crafted representations, followed by an examination of contemporary approaches based on convolutional neural networks (CNNs) that have achieved significant performance. In addition, the most frequently utilized datasets are addressed, and lastly, promising research routes in this rapidly increasing area are indicated.

Keywords: Crowd counting, density estimation, crowd analysis.

8.1 Introduction

Nowadays, the overall growth of the world's population requires careful planning in order to prevent crowd congestion [1]. Although in recent years the management and surveillance of events has attracted a great deal of attention, defining the exact number of individuals in places such as airports and/or commercial areas and monitoring their contact situations, has also become crucially important for global health due to the COVID-19 pandemic that has taken place in the past two years [2]. This area aims to analyze the individuals' behavior belonging to a large group, extracting crucial information from video in which large crowds are present. In these scenarios, crowd counting plays a necessary role for social safety and control management. Crowd density analysis evaluates the number of subjects in a crowded place, thus generating a corresponding density estimation map representing the number of people per pixel, as shown in Figure 8.1.

First of all, it is essential to establish a general definition for crowd behavior analysis and characterization. In [3], the authors define the concept of "crowd" by differentiating it from the definition of a "group." A *Group* constitutes a people series whose number can vary from two to one hundred [4, 5]. A *Crowd* (or mass) is represented by a single wide subject nucleus whose members have a common physical localization [6]. Therefore, a crowd usually relates to a much higher set of individuals whose relationships are less strong and with an organization that issues from the one-man interplay between agents. Once the notion of a crowd is correctly determined, circumscribing the scope of the area is the successful defiance. There are very clear clues related to crowd behavior analysis. Generally, one of the principal purposes is to interpret the behavior belonging to a high individual concentration through the information captured chiefly from video sources [7]. Additionally, there are various aspects that may be important to uncover. For instance, knowing the total number of people in a given location and how this number changes over a fixed time period could be helpful in preventing hazardous pedestrian stampedes. Another example could be the analysis of

(a)

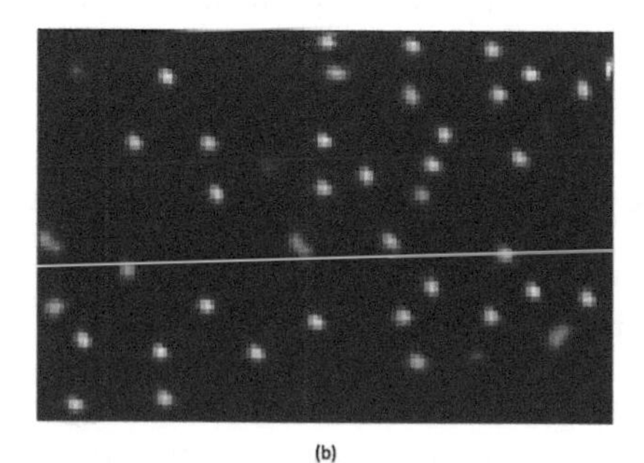

(b)

Figure 8.1 (a) Input sample image, (b) corresponding density map.

Figure 8.2 Different crowded scenes.

the predominant motion directions, which is particularly suitable for detecting individuals whose motions defer from the main flows, thus identifying the reasons.

Clutter, occlusions, uneven distribution of people, lighting, intra-and inter-scene transitions, staircases, and prospects all contribute to crowd density estimation being an extremely difficult task. Figure 8.2 illustrates some of these challenges. In recent years, the high complexity related to the various applications has meant that there has been greater consideration by researchers. Recent systematic reviews of crowd analysis [8–11], identify several critical applications (see Figure 8.3).

Security Monitoring: The use of video surveillance cameras in public places such as airports, stadiums, and tourist resorts, has allowed for easier crowd analysis. Despite this, traditional algorithms may not work in dense crowds due to development limitations [12, 13].

Disaster Management: Many situations embroiling crowd gatherings could lead to disasters such as stampedes. In these cases, crowd monitoring can be useful as a tool for early detection of overcrowding, thus avoiding aversion to any disaster [14].

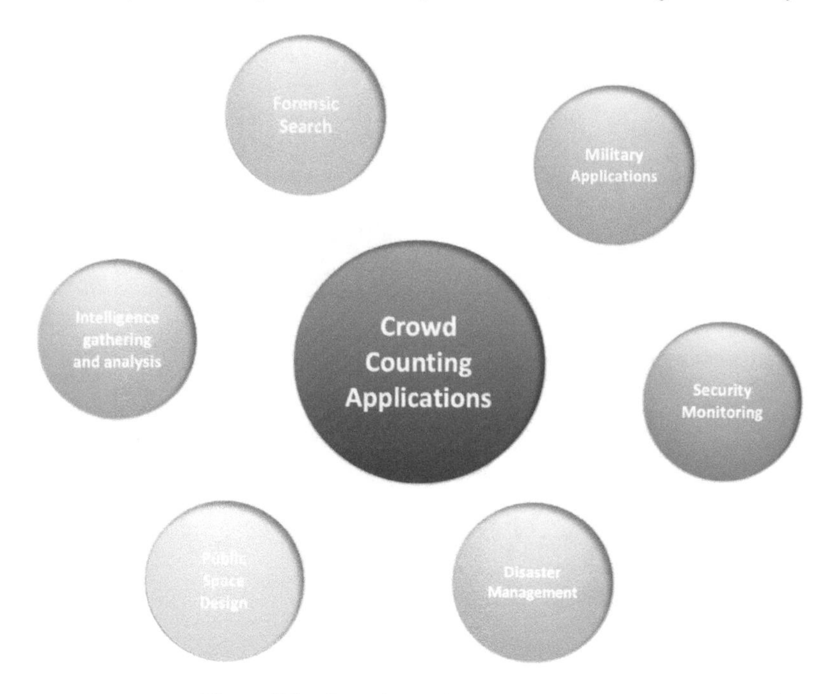

Figure 8.3 Crowd counting application fields.

Public Space Design: Crowd analysis can be very helpful in designing public areas that are optimized to improve security and motion [15].

Intelligence gathering and analysis: Crowd counting methods are useful for gathering information for further analysis. For example, crowd density could be useful to assess people's interest in a commercial product, analyze the flow of pedestrians over multiple days, thus optimizing signal waiting times, etc [16].

Forensic Search: Crowd analysis is of crucial importance in bombing occurrences, shootings, or incidents in large rallies, especially for the suspects and victims' search. Classical biometric algorithms for detecting and recognizing individuals can be enhanced by adopting crowd analysis techniques [17].

Military Applications: The fighter jets, soldiers, and drones number are evaluated by means of appropriate crowd analysis systems [18].

This variety of applications has led research workers from different fields to design innovative technologies while also analyzing various topics related to crowd analysis, i.e., counting, density estimation, and behavior analysis.

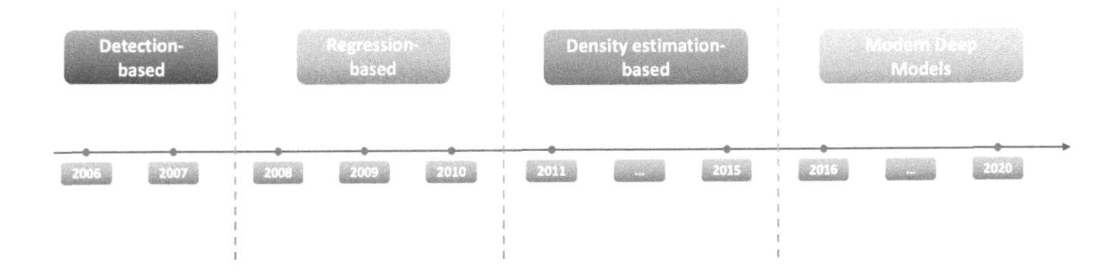

Figure 8.4 A brief timeline of important crowd counting milestones. Deep learning techniques are first introduced in 2015. See Sections 8.2 and 8.3 for more detailed description.

With recent progress in deep learning, and in particular convolutional neural networks (CNN) architectures, the crowd counting field reached meaningful advancements. Traditional models usually exploit basic CNNs, achieving considerable advancement with respect to heuristic methods. Recently, the models based on the fully convolutional network (FCN) represent the architecture par excellence. A brief timeline is illustrated in Figure 8.4, which includes the principal milestones of crowd density estimation techniques.

The article organization is as follows: Section 8.2 shortly summarizes the traditional crowd density methods. Section 8.3 describes several CNN-based approaches present in literature in accordance with the network properties and the inference algorithm method. Section 8.4 provides an overview of the most frequently used datasets and the standard metrics adopted to compare the different cutting-edge methods. Section 8.5 highlights several points about future trends and open challenges and, at last, concluding remarks are present in Section 8.6.

8.2 Traditional Crowd Counting Approaches

It is possible to distinguish several approaches in literature to address the crowd counting problem in images and/or videos. These methods are principally classified into two classes: supervised and unsupervised crowd density. In the first approach (i.e., supervised), it is possible to build a model starting from labeled data, with which predictions are obtained. In unsupervised learning, unstructured data is available. These two classes are further separated into the following types: detection-based, regression-based, traditional density estimation and convolutional neural network (CNN)-based. On the other hand, unsupervised techniques include counting by clustering [19, 20]. The categorization of crowd counting techniques is

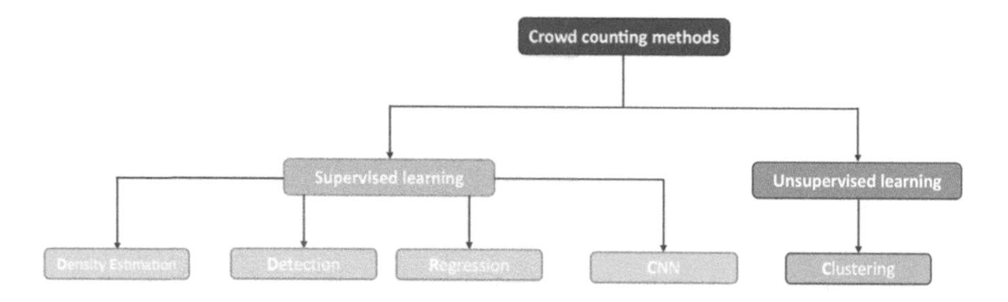

Figure 8.5 Crowd counting techniques in literature.

illustrated in Figure 8.5. The main objective is to analyze the recent CNN architectures explored in the crowd density field while also reviewing the most significant methodologies, as described below.

8.2.1 Detection-based approaches

Initially, the researchers focused their attention by adopting a detection framework. In more detail, given an image and/or a video sequence, it is possible to use a sliding window to detect the main parts of an individual, subsequently applying various hand-crafted features [26]. Recently, several CNNs-based object detectors, such as region-based CNN (R-CNN) [26], faster R-CNN [22], and mask R-CNN [23] have been experimented with, achieving higher detection performance than classic methods. However, such approaches are not cutting-edge as they do not perform satisfactorily and they are not competitive in presence of occlusions and crowded scenes.

8.2.2 Regression-based approaches

To overcome the limitations that characterize the above-mentioned methods, several researchers tackle crowd counting as a regression problem, mapping the features extracted from the image to corresponding object density maps. These methods consist of two phases: feature extraction and regression modeling. Several global and local descriptors are applied for encoding object information, such as edge, texture and gradient features, SIFT [24], HOG, LBP, and GLCM [25]. After that, various regression models—Linear, Gaussian process and Ridge [26]—can be utilized to learn mapping among the current crowd counting and low-level characteristics.

8.2.3 Density estimation approaches

Although regression-based methods performed very well in presence of occlusions and disorder, many of them overlooked significant spatial features as they were regressing in the overall count.

Using this approach, an estimate of an unobservable probability-density function is obtained through observed data, thus using spatial information. In this regard, in [27], the authors incorporated spatial characteristics into the learning process by proposing a linear mapping between local information and the corresponding object density maps. By observing the difficulty in the linear learning process, several studies have introduced non-linear mapping. Density estimation (DE) algorithm approaches can be classified into three levels [9]:

- Low-level DE methods: these techniques are based on motion characteristics, i.e., background segmentation, tracking methods, etc.
- Middle-level DE methods: patterns in data depend on classification algorithms.
- High-level density estimation methods: dynamic texture approaches are included in this group.

8.2.4 Clustering approaches

Clustering is defined as a set of multivariate data analysis methods aimed at selecting and grouping homogeneous elements in a dataset. Crowd counting approaches by clustering are founded on measures relating to the similarity among elements (in this specific case, visual and movement characteristics). For example, in [28] the authors adopt the Kanade–Lucas–Tomasi (KLT) feature tracker combined with Bayesian clustering to approximate the people number in a frame. Since these methods base their operation exclusively on appearance features, very often there is an erroneous estimate when the subjects are static. Consequently, these approaches achieve discrete results by analyzing continuous image frames.

8.3 CNN-based Approaches

Advances by CNNs in manifold computer vision tasks have allowed researchers to focus their efforts on the use of these recent techniques. Several CNN-based approaches have been introduced in literature, which can be classified according to the network properties and the inference methodology.

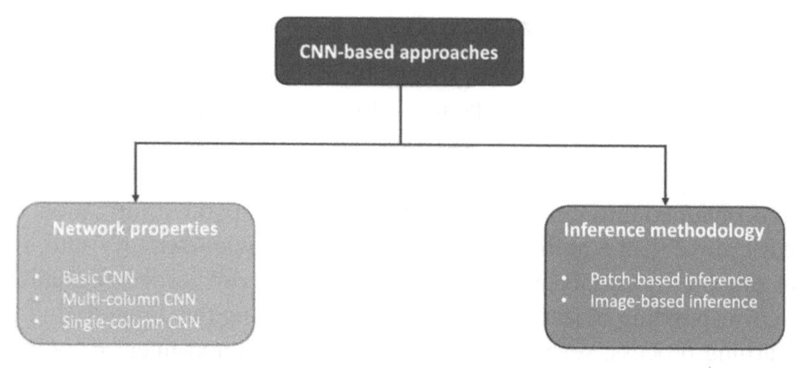

Figure 8.6 Classification of CNN-based approaches.

This classification is illustrated in Figure 8.6. In more detail, the following categories can be distinguished:

- **Basic CNN**: mainly includes early approaches developed for crowd counting. They usually consist of three types of layers, which are convolutional layers, pooling layers, and fully-connected layers.
- **Multi-column CNN**: it consists of different columns to capture multi-scale information, thus generating excellent quality crowd density maps.
- **Single-column CNN**: features deeper end-to-end CNNs to explore critical situations in crowd counting, thus producing high-quality density maps.

These approaches can be further divided according to the different training manners:

- **Patch-based inference**: CNNs are performed via random frames from the input image. To get the predictions, a sliding window is applied over the whole picture. The results are calculated for every window, finally summing them up to get the overall count.
- **Image-based inference**: these approaches work on the whole image, thus exploiting global information. However, the success of such methods is closely related to the frame resolution.

8.3.1 Basic CNN

Fu *et al.* [29] introduced for the first time the basic CNNs for crowd counting based on the multi-stage ConvNet. Soon after, Wang *et al.* [30] developed a deep CNN regression model. In [31], the authors propose a CNN-supervised learning framework while also tuning several state-of-the-art CNN architectures in the regression model. Elad *et al.* [32] employed CNNs, incorporating two meaningful advances: layered boosting and selective sampling. The basic CNN-based approaches mentioned above are simple and easily developed. However, their results are often restricted owing to the extrapolated information quality.

8.3.2 Multiple-column CNN

To this end, Zhang *et al.* [33] proposed a multi-column CNN which can estimate crowd numbers accurately in a single image. In [34], the authors adopted a combination of deep and shallow, fully convolutional networks, while in [35], a switching CNN that exploits intra-image crowd density variation was proposed. In order to address the pixels' interdependence, Zhang *et al.* [36] developed a relational attention network (RANet) to capture both long- and short-range pixel interdependence. To overcome the counting performance differences in several regions, two networks were introduced in [37], respectively, density attention network (DANet) and attention scaling network (ASNet). Recently, the authors in [38], presented an alternative approach to loosely supervised crowd counting through a sequence-to-count prediction method closely related to the transformer-encoder, namely TransCrowd. Sun *et al.* [39] developed a robust framework adopting transformers, thus encoding characteristics with global receptive fields, introducing the following modules: TAM stands for token-attention module, and RTM stands for regression-token module.

Despite the improvements found and the high results achieved by M-C architectures, they still suffer from different limitations. Among the best known, it is worth remembering the large computational costs closely related to training and redundant functionalities. These limitations have led several researchers to introduce single-column CNN, a very easy but skillful model to overcome these disadvantages and address several demanding scenarios.

8.3.3 Single-column CNN

In recent years, single-column CNN approaches have been the best solution for achieving meaningful performance, thanks above all to their simplicity and effective training process. Li *et al.* [40] designed a novel architecture, i.e., CSRNet, taking advantage of the dilated convolutional layers. In [41], a scale-adaptive CNN (SaCNN) framework with a backbone of fixed small receptive fields was proposed. Cao *et al.* [42] introduced an encoder–decoder

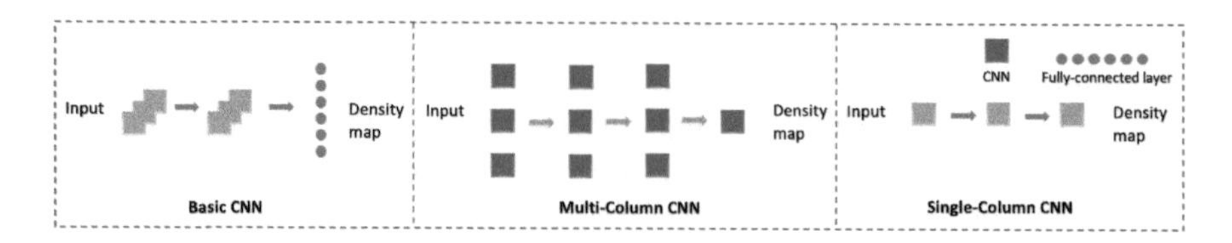

Figure 8.7 The three CNN-based network architectures.

network, namely scale aggregation network (SANet). In more detail, through the encoder, it is possible to extrapolate multi-scale information while the decoder has the task of creating high resolution density maps by means of transposed convolutions. A new trellis encoder–decoder architecture (TEDnet) for crowd counting was designed in [43].

Recently, attention-based methods have achieved meaningful results. In particular, instead of analyzing the entire image, through this mechanism it is possible to concentrate only on the most important areas. In this regard, Liu *et al.* [44] developed ADCrowdNet, an attention-based architecture that includes two concatenated networks: an attention map generator and a density map estimator. The three CNN-based network architectures are shown in Figure 8.7.

8.3.4 Patch-based and image-based inference

In patch-based inference approaches, CNNs are performed via random frames from the input image. To get the predictions, a sliding window is applied over the whole picture. The results are calculated for every window, finally summing them up to get the total count. Zhang *et al.* [45] designed a CNN model for crowd counting, training it on casual frames extracted from the original input image. As a result, the object count in each frame is calculated by supplementing the density map. Another patch-based inference approach is available in [46], where the authors developed an image patch rescaling module (PRM).

Compared to the above-mentioned approach, in image-based inference it is possible to work on the whole image, thus exploiting global information. However, the success of such methods is closely related to the frame resolution. In this regard, in [47], an end-to-end CNN architecture that makes use of a recurrent neural network with memory cells was introduced, taking the overall frame as input and straight producing the counting as output.

8.4 Benchmark Datasets and Evaluation Metrics

In recent years, several datasets have been created that stimulate researchers to create methods with better generalization capabilities. In fact, while older databases include low-density crowd images, newer ones analyze high-density crowds, scale changes, clutter and severe occlusion. This section provides an overview of the most frequently used datasets [48], including sample images in Figure 8.8 and a brief summary in Table 8.1. Finally, the standard metrics adopted to compare the different state-of-the-art methods are described.

8.4.1 Databases

UCSD dataset: The UCSD benchmark [49] Figure 8.8 (a) represents the first images collection originally created to count people. It is made up of 2000 frames, size 238×158, acquired by a video camera with ground truth labeling of all passersby in any fifth frame, for a total of 49,885 pedestrian instances.

Mall dataset: The Mall dataset Figure 8.8 (b) [50] includes videos captured via a security camera in a shopping center. Each video is characterized by 2000 frames, dimension 320×240 with 62,325 pedestrians labeled overall. It is characterized by various density levels and several activity patterns.

UCF_CC_50 dataset: The UCF_CC_50 database Figure 8.8 (c) [51] is the first really demanding benchmark with a large gamma of different densities and scenes. It includes publicly available web images and the tags are varied, such as protests, concerts, stadiums, and marathons. A total of 63,075 subjects were annotated.

WorldExpo'10 dataset: The purpose of the WorldExpo'10 dataset Figure 8.8 (d) [52] creation is the analysis of cross-scene crowd counting. The databases comprise 1132 annotated video sequences (3980 pictures of dimension 576×720, 199,923 labeled pedestrians) acquired by video surveillance cameras, all from the Shanghai 2010 World Expo event.

Shanghai Tech dataset: The Shanghai Tech benchmark Figure 8.8 (e,f) [33] represents one of the largest large-scale crowd density databases, including 1198 pictures with 330,165 labeled heads. At the state-of-the-art, Shanghai Tech is one of the greatest DBs by the number of labeled topics and is composed by two parts: Part A and Part B. Part A compounds causally selected frames from the web, while Part B includes pictures captured from metropolitan areas of Shanghai.

Figure 8.8 Sample images from the most frequently used crowd counting datasets.

UCF-QNRF dataset: The UCF-QNRF Figure 8.8 (g) [53], compared to the latest generation databases, represents the greatest benchmark adopted for several topics and has a wide kind of scenes including different viewpoints, densities and illumination variations. The images are acquired from the web,

Table 8.1 Brief summary of the most frequently used datasets, including crowd counting and other statistical information.

Datasets	N. images	Size	Max	Min	Avg.	Total
UCSD [49]	2000	238×158	46	11	25	49,885
Mall [50]	2000	320×240	53	13	31	62,325
UCF_CC_50 [51]	50	2101×2888	4,543	94	1,279	63,974
WorldExpo'10 [52]	3980	576×720	253	1	50	199,923
Shanghai Tech A [33]	482	589×868	3,139	33	501	241,677
Shanghai Tech B [33]	716	768×1024	578	9	123	88,488
UCF-QNRF [53]	1535	2013×2902	12,865	49	815	1,251,642
GCC [54]	15,212	1080×1920	3,995	0	501	7,625,843

including Hajj footage and Flickr platform pictures. It includes very realistic scenes extrapolated in uncontrolled conditions.

GCC dataset: The GCC dataset Figure 8.8 (h) [54] is created from an electronic game namely "GTA5 Crowd Counting" and comprises 15,212, with dimensions 1080×1920, for a total of 7,625,843 subjects. It features many benefits, such as high-resolution and so many different scenes with accurate labels.

8.4.2 Metrics

The most common performance evaluation metrics are the mean absolute error (MAE) and the mean square error (MSE). The MAE represents the average absolute deviation between the estimates and the actual observations. The MSE calculates the squared difference between the effective and expected output for each sample; its root is defined as the root mean squared error (RMSE).

They are formulated by the following equations, where n indicates the test set images number, t'_j is the forecast calculated and t_j is the ground truth.

$$MAE = \frac{1}{n}\sum_{j=1}^{n} |t_j - t\prime_j| \tag{8.1}$$

$$RMSE = \sqrt{\frac{1}{n}\sum_{j=1}^{n}(t_j - t\prime_j)^2} \tag{8.2}$$

Other common metrics used to compare recent studies and assess the efficiency of the proposed approach are the structural similarity index (SSIM) and peak signal to noise ratio (PSNR), in order to estimate the generated density map quality. In general, PSNR is characterized by the error between corresponding pixels. The calculation formula is shown as follows

$$PSNR = 20 * log_{10} \frac{MAX_I}{\sqrt{MSE}}, \tag{8.3}$$

where MAX_I represents the image's maximum possible pixel value. On the other hand, SSIM analyzes the images' similarity with respect to brightness, contrast, and structure. See the following equation:

$$SSIM = \frac{(2\mu_x\mu_y + c_1)(2\delta_{xy} + c_2)}{(\mu_x^2 + \mu_y^2 + c_1)(\delta_x^2 + \delta_y^2 + c_2)}, \tag{8.4}$$

where μ_x, δ_x^2, μ_y, and δ_y^2 indicate, respectively, the average and the variance of x and y. δ_{xy} is the covariance of x and y. c_1 and c_2 are two variables adopted to stabilize the division with weak denominator and can be defined as $c_1 = (k_1, L)^2$ and $c_2 = (k_2, L)^2$, where L is the dynamic range of the pixel values, $k_1 = 0.01$ and $k_2 = 0.03$.

Finally, average precision (AP) and average recall (AR) measure model localization performance.

A performance evaluation of the most representative algorithms discussed in Section 8.3 in terms of MAE and RMSE are shown in Table 8.2.

8.5 Open Challenges and Future Directions

Based on the analysis performed, several points can be highlighted about future trends and open challenges in crowd analysis field. First, training deep networks requires the collection of large databases. Although there are numerous datasets in literature, only a few are aimed at large crowds. Another difficulty encountered consists of training deep networks on new scenes; consequently, the idea of transfer learning could be a very promising research area.

As shown in Figure 8.9, scale variation still represents an exacting task. Depending on the scenes, the subject numbers could vary greatly, forming a dense crowd or, conversely, situations in which individuals are scattered.

Such differences would result in a considerable imbalance for the model. To overcome this limitation, as described in Section 8.3.2, several studies use the fusion of multi-scale features or density maps of varying levels, employing multi-column architectures with varying receptive fields. Occlusions, uneven distributions, and perspective variations are also currently being studied by researchers.

Table 8.2 Comparison results on the most frequently used databases.

Dataset	Approach	MAE	RMSE
UCSD	Traditional [27]	1.59	-
	Multiple-column [33]	1.07	1.35
	Multiple-column [45]	1.60	3.31
	Multiple-column [35]	1.62	2.10
	Single-column [44]	1.09	1.35
UCF_CC_50	Traditional [24]	468	590.3
	Multiple-column [33]	377.6	509.1
	Multiple-column [45]	467.0	498.5
	Multiple-column [35]	318.1	439.2
	Multiple-column [36]	239.8	319.4
	Multiple-column [37]	268.3	373.2
	Single-column [40]	266.1	397.5
	Single-column [41]	314.9	424.8
	Single-column [42]	258.4	334.9
	Single-column [44]	273.6	362.0
WorldExpo'10	Multiple-column [33]	11.6	-
	Multiple-column [45]	12.9	-
	Multiple-column [35]	9.4	-
	Single-column [41]	8.5	-
	Single-column [44]	7.3	-
Shanghai Tech A	Multiple-column [33]	110.2	173.2
	Multiple-column [45]	181.8	277.7
	Multiple-column [35]	90.4	135
	Multiple-column [36]	59.4	102.0
	Multiple-column [37]	71.4	120.6
	Single-column [40]	68.2	115
	Single-column [41]	86.8	139.2
	Single-column [42]	67.0	104.5
	Single-column [44]	70.9	115.2
Shanghai Tech B	Multiple-column [33]	26.4	41.3
	Multiple-column [45]	32.0	49.8
	Multiple-column [35]	21.6	33.4
	Multiple-column [36]	7.9	12.9
	Multiple-column [37]	9.1	14.7
	Single-column [40]	10.6	16
	Single-column [41]	16.2	25.8
	Single-column [42]	8.4	13.6
	Single-column [44]	7.7	12.9
UCF-QNRF	Multiple-column [35]	228	445
	Multiple-column [36]	111	190

Figure 8.9 Some challenges in crowd counting.

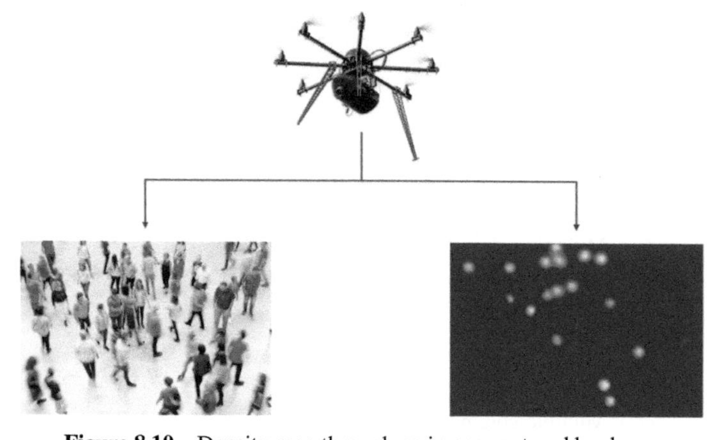

Figure 8.10 Density map through an image captured by drone.

Most of the methods developed so far have mostly been designed around single images. Very interesting could be the development of new methodologies aimed at operating in real time to count people from videos [55, 56]. In this regard, research is moving in new directions, such as the adoption of unmanned aerial vehicles (UAVs). Several studies are leveraging the recent CNN-based architectures developed to generate spatial density maps through aerial images captured by drones [57, 58] (See Figure 8.10). In light of this, it is believed that this field will have numerous advances.

8.6 Conclusion

This work presents a brief overview of recent methodologies in crowd counting analysis. The traditional approaches present in literature were initially summarized and, subsequently, the focus of the description was characterized by the recent advances in deep learning-based approaches. As it is not possible to cover all the crowd counting literature, a significant subset of the latest techniques was chosen for the analysis and review. In addition, the most frequently used databases and the common metrics to assess the crowd density models were described. Finally, a performance evaluation of the most representative algorithms was included, as well as identifying some of the most demanding challenges and issues.

References

[1] Beibei Zhan, Dorothy N Monekosso, Paolo Remagnino, Sergio A Velastin, and Li-Qun Xu. Crowd analysis: a survey. *Machine Vision and Applications*, 19(5):345–357, 2008.

[2] Osama S Faragallah, Sultan S Alshamrani, Heba M El-Hoseny, Mohammed A AlZain, Emad Sami Jaha, and Hala S El-Sayed. Utilization of deep learning-based crowd analysis for safety surveillance and spread control of covid-19 pandemic. *Intelligent Automation and Soft Computing*, pages 1483–1497, 2022.

[3] Francisco Luque Sánchez, Isabelle Hupont, Siham Tabik, and Francisco Herrera. Revisiting crowd behaviour analysis through deep learning: Taxonomy, anomaly detection, crowd emotions, datasets, opportunities and prospects. *Information Fusion*, 2020.

[4] Hayley Hung and Daniel Gatica-Perez. Estimating cohesion in small groups using audio-visual nonverbal behavior. *IEEE Transactions on Multimedia*, 12(6):563–575, 2010.

[5] Omar Adair Islas Ramírez, Giovanna Varni, Mihai Andries, Mohamed Chetouani, and Raja Chatila. Modeling the dynamics of individual behaviors for group detection in crowds using low-level features. In *2016 25th IEEE International Symposium on Robot and Human Interactive Communication (RO-MAN)*, pages 1104–1111. IEEE, 2016.

[6] HY Swathi, G Shivakumar, and HS Mohana. Crowd behavior analysis: A survey. In *2017 international conference on recent advances in electronics and communication technology (ICRAECT)*, pages 169–178. IEEE, 2017.

[7] Mounir Bendali-Braham, Jonathan Weber, Germain Forestier, Lhassane Idoumghar, and Pierre-Alain Muller. Recent trends in crowd analysis: A review. *Machine Learning with Applications*, page 100023, 2021.

[8] Vishwanath A Sindagi and Vishal M Patel. A survey of recent advances in cnn-based single image crowd counting and density estimation. *Pattern Recognition Letters*, 107:3–16, 2018.

[9] Khalil Khan, Waleed Albattah, Rehan Ullah Khan, Ali Mustafa Qamar, and Durre Nayab. Advances and trends in real time visual crowd analysis. *Sensors*, 20(18):5073, 2020.

[10] Bo Li, Hongbo Huang, Ang Zhang, Peiwen Liu, and Cheng Liu. Approaches on crowd counting and density estimation: a review. *Pattern Analysis and Applications*, pages 1–22, 2021.

[11] Zizhu Fan, Hong Zhang, Zheng Zhang, Guangming Lu, Yudong Zhang, and Yaowei Wang. A survey of crowd counting and density estimation based on convolutional neural network. *Neurocomputing*, 2021.

[12] Saiyed Umer, Mrinmoy Ghorai, and Partha Pratim Mohanta. Event recognition in unconstrained video using multi-scale deep spatial features. In *2017 Ninth International Conference on Advances in Pattern Recognition (ICAPR)*, pages 1–6, 2017.

[13] Khosro Rezaee, Sara Mohammad Rezakhani, Mohammad R Khosravi, and Mohammad Kazem Moghimi. A survey on deep learning-based real-time crowd anomaly detection for secure distributed video surveillance. *Personal and Ubiquitous Computing*, pages 1–17, 2021.

[14] Liu Bai, Cheng Wu, Feng Xie, and Yiming Wang. Crowd density detection method based on crowd gathering mode and multi-column convolutional neural network. *Image and Vision Computing*, 105:104084, 2021.

[15] Maha Hamdan Alotibi, Salma Kammoun Jarraya, Manar Salamah Ali, and Kawthar Moria. Cnn-based crowd counting through iot: Application for saudi public places. *Procedia Computer Science*, 163:134–144, 2019.

[16] Hyuncheol Kim, Jaeho Han, and Soonhung Han. Analysis of evacuation simulation considering crowd density and the effect of a fallen person. *Journal of Ambient Intelligence and Humanized Computing*, 10(12):4869–4879, 2019.

[17] Bryan Kneis. Face detection for crowd analysis using deep convolutional neural networks. In *International Conference on Engineering Applications of Neural Networks*, pages 71–80. Springer, 2018.

[18] Adrian Albert, Jasleen Kaur, and Marta C Gonzalez. Using convolutional networks and satellite imagery to identify patterns in urban environments at a large scale. In *Proceedings of the 23rd ACM SIGKDD international conference on knowledge discovery and data mining*, pages 1357–1366, 2017.

[19] Naveed Ilyas, Ahsan Shahzad, and Kiseon Kim. Convolutional-neural network-based image crowd counting: Review, categorization, analysis, and performance evaluation. *Sensors*, 20(1):43, 2020.

[20] Rafik Gouiaa, Moulay A Akhloufi, and Mozhdeh Shahbazi. Advances in convolution neural networks based crowd counting and density estimation. *Big Data and Cognitive Computing*, 5(4):50, 2021.

[21] Ross Girshick, Jeff Donahue, Trevor Darrell, and Jitendra Malik. Rich feature hierarchies for accurate object detection and semantic segmentation. In *Proceedings of the IEEE conference on computer vision and pattern recognition*, pages 580–587, 2014.

[22] Shaoqing Ren, Kaiming He, Ross Girshick, and Jian Sun. Faster r-cnn: Towards real-time object detection with region proposal networks. *Advances in neural information processing systems*, 28:91–99, 2015.

[23] Kaiming He, Georgia Gkioxari, Piotr Dollár, and Ross Girshick. Mask r-cnn. In *Proceedings of the IEEE international conference on computer vision*, pages 2961–2969, 2017.

[24] Haroon Idrees, Imran Saleemi, Cody Seibert, and Mubarak Shah. Multi-source multi-scale counting in extremely dense crowd images. In *Proceedings of the IEEE conference on computer vision and pattern recognition*, pages 2547–2554, 2013.

[25] Jessie R Balbin, Ramon G Garcia, Kaira Emi D Fernandez, Nicolo Paolo G Golosinda, Karyl Denise G Magpayo, and Robee Jasper B Velasco. Crowd counting system by facial recognition using histogram of oriented gradients, completed local binary pattern, gray-level co-occurrence matrix and unmanned aerial vehicle. In *Third International Workshop on Pattern Recognition*, volume 10828, page 108280Y. International Society for Optics and Photonics, 2018.

[26] Antoni B Chan and Nuno Vasconcelos. Counting people with low-level features and bayesian regression. *IEEE Transactions on image processing*, 21(4):2160–2177, 2011.

[27] Victor Lempitsky and Andrew Zisserman. Learning to count objects in images. *Advances in neural information processing systems*, 23:1324–1332, 2010.

[28] Vincent Rabaud and Serge Belongie. Counting crowded moving objects. In *2006 IEEE Computer Society Conference on Computer Vision and Pattern Recognition (CVPR'06)*, volume 1, pages 705–711. IEEE, 2006.

[29] Min Fu, Pei Xu, Xudong Li, Qihe Liu, Mao Ye, and Ce Zhu. Fast crowd density estimation with convolutional neural networks. *Engineering Applications of Artificial Intelligence*, 43:81–88, 2015.

[30] Chuan Wang, Hua Zhang, Liang Yang, Si Liu, and Xiaochun Cao. Deep people counting in extremely dense crowds. In *Proceedings of the 23rd ACM International Conference on Multimedia*, MM '15, page 1299–1302, New York, NY, USA, 2015. Association for Computing Machinery.

[31] Yao Xue, Nilanjan Ray, Judith Hugh, and Gilbert Bigras. Cell counting by regression using convolutional neural network. In *European Conference on Computer Vision*, pages 274–290. Springer, 2016.

[32] Elad Walach and Lior Wolf. Learning to count with cnn boosting. In *European conference on computer vision*, pages 660–676. Springer, 2016.

[33] Yingying Zhang, Desen Zhou, Siqin Chen, Shenghua Gao, and Yi Ma. Single-image crowd counting via multi-column convolutional neural network. In *Proceedings of the IEEE conference on computer vision and pattern recognition*, pages 589–597, 2016.

[34] Lokesh Boominathan, Srinivas SS Kruthiventi, and R Venkatesh Babu. Crowdnet: A deep convolutional network for dense crowd counting. In *Proceedings of the 24th ACM international conference on Multimedia*, pages 640–644, 2016.

[35] Deepak Babu Sam, Shiv Surya, and R Venkatesh Babu. Switching convolutional neural network for crowd counting. In *Proceedings of the IEEE conference on computer vision and pattern recognition*, pages 5744–5752, 2017.

[36] Anran Zhang, Jiayi Shen, Zehao Xiao, Fan Zhu, Xiantong Zhen, Xianbin Cao, and Ling Shao. Relational attention network for crowd counting. In *Proceedings of the IEEE/CVF International Conference on Computer Vision*, pages 6788–6797, 2019.

[37] Xiaoheng Jiang, Li Zhang, Mingliang Xu, Tianzhu Zhang, Pei Lv, Bing Zhou, Xin Yang, and Yanwei Pang. Attention scaling for crowd counting. In *Proceedings of the IEEE/CVF Conference on Computer Vision and Pattern Recognition*, pages 4706–4715, 2020.

[38] Dingkang Liang, Xiwu Chen, Wei Xu, Yu Zhou, and Xiang Bai. Transcrowd: Weakly-supervised crowd counting with transformer. *arXiv preprint arXiv:2104.09116*, 2021.

[39] Guolei Sun, Yun Liu, Thomas Probst, Danda Pani Paudel, Nikola Popovic, and Luc Van Gool. Boosting crowd counting with transformers. *arXiv preprint arXiv:2105.10926*, 2021.

[40] Yuhong Li, Xiaofan Zhang, and Deming Chen. Csrnet: Dilated convolutional neural networks for understanding the highly congested

scenes. In *Proceedings of the IEEE conference on computer vision and pattern recognition*, pages 1091–1100, 2018.

[41] Lu Zhang, Miaojing Shi, and Qiaobo Chen. Crowd counting via scale-adaptive convolutional neural network. In *2018 IEEE Winter Conference on Applications of Computer Vision (WACV)*, pages 1113–1121. IEEE, 2018.

[42] Xinkun Cao, Zhipeng Wang, Yanyun Zhao, and Fei Su. Scale aggregation network for accurate and efficient crowd counting. In *Proceedings of the European Conference on Computer Vision (ECCV)*, pages 734–750, 2018.

[43] Xiaolong Jiang, Zehao Xiao, Baochang Zhang, Xiantong Zhen, Xianbin Cao, David Doermann, and Ling Shao. Crowd counting and density estimation by trellis encoder-decoder networks. In *Proceedings of the IEEE/CVF Conference on Computer Vision and Pattern Recognition*, pages 6133–6142, 2019.

[44] Ning Liu, Yongchao Long, Changqing Zou, Qun Niu, Li Pan, and Hefeng Wu. Adcrowdnet: An attention-injective deformable convolutional network for crowd understanding. In *Proceedings of the IEEE/CVF Conference on Computer Vision and Pattern Recognition*, pages 3225–3234, 2019.

[45] Cong Zhang, Hongsheng Li, Xiaogang Wang, and Xiaokang Yang. Cross-scene crowd counting via deep convolutional neural networks. In *Proceedings of the IEEE conference on computer vision and pattern recognition*, pages 833–841, 2015.

[46] Usman Sajid and Guanghui Wang. Plug-and-play rescaling based crowd counting in static images. In *Proceedings of the IEEE/CVF Winter Conference on Applications of Computer Vision*, pages 2287–2296, 2020.

[47] Chong Shang, Haizhou Ai, and Bo Bai. End-to-end crowd counting via joint learning local and global count. In *2016 IEEE International Conference on Image Processing (ICIP)*, pages 1215–1219. IEEE, 2016.

[48] Guangshuai Gao, Junyu Gao, Qingjie Liu, Qi Wang, and Yunhong Wang. Cnn-based density estimation and crowd counting: A survey. *arXiv preprint arXiv:2003.12783*, 2020.

[49] Antoni B Chan, Zhang-Sheng John Liang, and Nuno Vasconcelos. Privacy preserving crowd monitoring: Counting people without people models or tracking. In *2008 IEEE Conference on Computer Vision and Pattern Recognition*, pages 1–7. IEEE, 2008.

[50] Ke Chen, Chen Change Loy, Shaogang Gong, and Tony Xiang. Feature mining for localised crowd counting. In *Bmvc*, volume 1, page 3, 2012.

[51] Haroon Idrees, Imran Saleemi, Cody Seibert, and Mubarak Shah. Multi-source multi-scale counting in extremely dense crowd images. In *Proceedings of the IEEE conference on computer vision and pattern recognition*, pages 2547–2554, 2013.

[52] Cong Zhang, Hongsheng Li, Xiaogang Wang, and Xiaokang Yang. Cross-scene crowd counting via deep convolutional neural networks. In *Proceedings of the IEEE conference on computer vision and pattern recognition*, pages 833–841, 2015.

[53] Haroon Idrees, Muhmmad Tayyab, Kishan Athrey, Dong Zhang, Somaya Al-Maadeed, Nasir Rajpoot, and Mubarak Shah. Composition loss for counting, density map estimation and localization in dense crowds. In *Proceedings of the European Conference on Computer Vision (ECCV)*, pages 532–546, 2018.

[54] Haroon Idrees, Muhmmad Tayyab, Kishan Athrey, Dong Zhang, Somaya Al-Maadeed, Nasir Rajpoot, and Mubarak Shah. Composition loss for counting, density map estimation and localization in dense crowds. In *Proceedings of the European Conference on Computer Vision (ECCV)*, pages 532–546, 2018.

[55] Feng Xiong, Xingjian Shi, and Dit-Yan Yeung. Spatiotemporal modeling for crowd counting in videos. In *Proceedings of the IEEE International Conference on Computer Vision*, pages 5151–5159, 2017.

[56] Habib Ullah, Ihtesham Ul Islam, Mohib Ullah, Muhammad Afaq, Sultan Daud Khan, and Javed Iqbal. Multi-feature-based crowd video modeling for visual event detection. *Multimedia Systems*, 27(4):589–597, 2021.

[57] Giovanna Castellano, Ciro Castiello, Marco Cianciotta, Corrado Mencar, and Gennaro Vessio. Multi-view convolutional network for crowd counting in drone-captured images. In *European Conference on Computer Vision*, pages 588–603. Springer, 2020.

[58] Zhihao Liu, Zhijian He, Lujia Wang, Wenguan Wang, Yixuan Yuan, Dingwen Zhang, Jinglin Zhang, Pengfei Zhu, Luc Van Gool, Junwei Han, et al. Visdrone-cc2021: The vision meets drone crowd counting challenge results. In *Proceedings of the IEEE/CVF International Conference on Computer Vision*, pages 2830–2838, 2021.

9

Recent Trends in 2D Object Detection and Applications in Video Event Recognition

Prithwish Jana[1] and Partha Pratim Mohanta[2]

[1]Dept. of Computer Science & Engg., Indian Institute of Technology, India
[2]Electronics & Communication Sc. Unit, Indian Statistical Institute, India
E-mail: pjana@ieee.org; partha.p.mohanta@gmail.com

Abstract

Object detection serves as a significant step in improving performance of complex downstream computer vision tasks. It has been extensively studied for many years now and current state-of-the-art 2D object detection techniques proffer superlative results even in complex images. In this chapter, we discuss the geometry-based pioneering works in object detection, followed by the recent breakthroughs that employ deep learning. Some of these use a monolithic architecture that takes a RGB image as input and passes it to a feed-forward ConvNet or vision Transformer. These methods, thereby predict class-probability and bounding-box coordinates, all in a single unified pipeline. Two-stage architectures on the other hand, first generate region proposals and then feed it to a CNN to extract features and predict object category and bounding-box. We also elaborate upon the applications of object detection in video event recognition, to achieve better fine-grained video classification performance. Further, we highlight recent datasets for 2D object detection both in images and videos, and present a comparative performance summary of various state-of-the-art object detection techniques.

Keywords: Object detection, activity classification, video event recognition, localization and classification, deep learning.

9.1 Introduction

Computer vision deals with mimicking the human visual perception system, such that computers can "see" (perceive and understand) their surrounding visual scenes the same way as we humans do. Through their perception subsystem, computers first perceive scenes through cameras, which are sent for internal processing commonly in the form of images and videos. One of the important tasks on images is *object detection*, and it may be regarded as a primary step in many computer vision pipelines to support downstream complex tasks like panoptic and instance segmentation, image

Figure 9.1 Digital images are captured by cameras through perception subsystem. The example image (from MS COCO dataset [1]) when fed to an object detection system, detects two objects, estimates tight bounding boxes around them and classifies the object within as "bear" and "bird."

captioning, etc. Object detection deals with identifying instances of certain entities (e.g., human, dog, chair, tree, etc.) within a digital image. As illustrated in Figure 9.1, identification is in terms of *localizing* a bounding box which encompasses an object and *classifying* the object present within the bounding box.

The task of video event recognition [3] is to predict the ongoing event or activity in a video throughout its duration. The prevalent approach is to obtain a global video-level feature representation [4] from 3D CNNs and classify the video using the same. However, this is generally insufficient for fine-grained tasks. As for example, it is a difficult feat [5] to discriminate similar events like anniversary event and birthday party, both of which showcase some kind of gathering of people and celebration. In such cases, frame-wise object detection may serve to be useful that can identify objects involved throughout the video duration. The overall idea is depicted in Figure 9.2. Event-specific

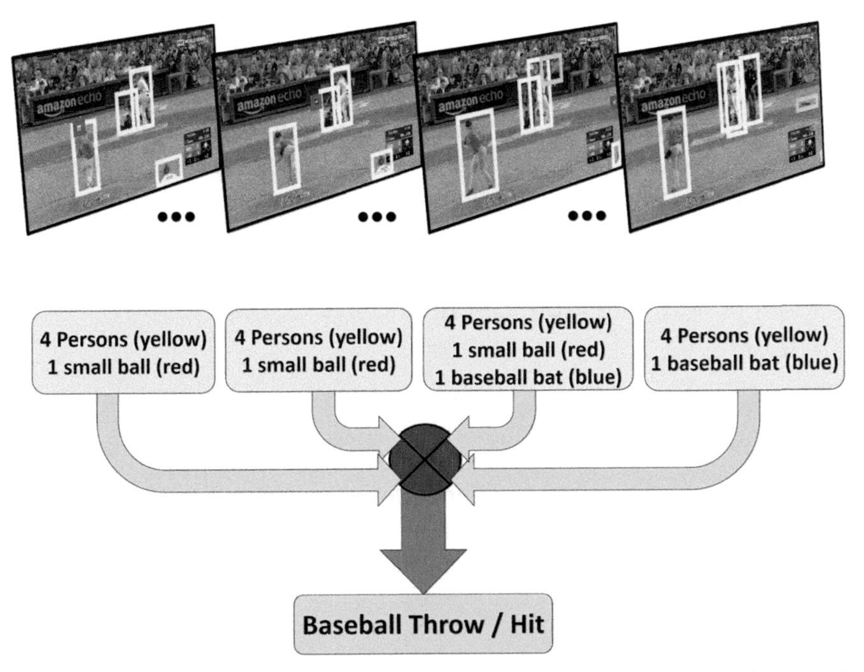

Figure 9.2 Object detection on each frame of a video gives an estimate of the relative position of each entity throughout the video duration. Efficient combination of this sequential information can lead to superlative event recognition in videos. For the example video [2], detected objects (person, baseball bat, ball) are shown per-frame, that are used to predict final event "Baseball Throw/Hit."

objects (e.g., balloons, toys) or event-specific inter-frame motion of objects (e.g., periodic round-about motion of persons can hint about the game of musical chairs) can be particularly useful in fine-grained discrimination of video events.

The rest of the chapter is organized as follows. In Section 9.2 of this chapter, we first highlight some of the early geometric approaches to object detection. We move on to describe some of the recent major advancements in object detection that employ deep learning approaches. Next, in Section 9.3, we elucidate on techniques of fine-grained video event recognition by frame-wise object detection. In Section 9.4, we present some of the 2D object recognition datasets both in the domain of images and videos. We compare some of the recent state-of-the-art object detection techniques in Section 9.5. Finally, we conclude in Section 9.6.

9.2 Related Work on 2D Object Detection

In this section, we discuss some of the early rule-based and geometry-based 2D object detection techniques. We further describe later progresses made by adopting deep learning-based architectures that employ either monolithic architectures (a single-stage object detection pipeline) and two-stage architectures (where region proposals are generated first, followed by classification, regression, and post-processing).

9.2.1 Early object detection: Geometrical and shape-based approaches

The earliest object detection techniques were majorly based on pattern or template matching. One such notable method was that by Fischler *et al.*'s [6] in 1973, who devised a technique of matching pictorial structures. The authors mention a commonality between different images of the same object, viz. they all can be broken down into a number of fundamental *structures* possessing some preset relative spatial position. As such, given a test image, if all such fundamental structures are present and they maintain the same relative position it can be said that the object is present within the image. They put forward an embedding metric to determine how well each of the fundamental structures make up the total composite object that is being probed and used sequential optimization for matching. Most of the object detection methods in that era exploited object shape and geometry to solve the detection problem. Such approaches were in general "hard-detection" techniques, where a composite shape and inter-structure configuration is evaluated only after all constituent structures could be recognized and detected in an image. In contrast to such hard-detection, Burl *et al.* [7] suggested a soft-detection technique to identify an object composed of several structures whose relative position may vary upto an extent. As a part of this, they probed that particular relative configuration of these structures which simultaneously maximized the composite shape's log-likelihood (global criteria) and matching score for the structures (local criteria). The scenario becomes somewhat different when images are not in RGB format, but is grayscale (i.e., one channel, e.g., infrared images used in military operations) or stacked (e.g., confocal microscopy images). Quy *et al.* [8] and Miao *et al.* [9] used binarization [10] to perform thresholding on the grayscale infrared images and compute bounding-box around the prominent-cum-salient connected components. Luo *et al.* [11] also use thresholding as a preprocessing step before fine-grained object detection in confocal laser-scanning microscope image stacks. As time progressed, the trend of shape-based template matching and rule-based methods for object detection shifted toward statistical approaches and machine learning-based classifiers.

9.2.2 Modern object detection techniques: Use of deep learning

We discuss some of the recent significant monolithic and two-stage architectures for 2D object detection, in the subsequent paragraphs.

9.2.2.1 Monolithic architectures for 2D object detection

The work on *DetectorNet* by Szegedy *et al.* [12] was one of the earliest that incorporated deep neural networks to predict category and precise location of objects within an image. Prior to that, deep convolutional neural networks (CNN) were mostly used for image classification tasks. On the contrary, DetectorNet was built upon the AlexNet [13] architecture which was specialized for the task of detection. This was done by replacing the last layer of softmax with a regression layer that outputs a square-shaped mask of fixed size. They considered input of receptive field 225×225 and output mask of size 24×24. Further, they dissect the image into multi-scale coarse grids and consider four supplementary networks for estimating the top, bottom, right, and left halves of objects in these grids. All these predicted masks on coarse grid help each other and the full mask in correcting mistakes and on combining and refinement, proffers the predicted bounding box for objects.

To train the network, $L2$-loss of the predicted mask and the ground-truth mask is minimized over all samples in the training set. Although this course-to-fine approach of object mask regression was a pioneer in CNN-based object detection, but there was a disadvantage. Since the whole CNN architecture needs to be trained separately for every object category and every mask type, this incurs huge computational complexity.

Another notable work published at around the same time was *OverFeat*, put forward by Sermanet *et al.* [14]. Similar to DetectorNet, this is also a monolithic shared architecture to simultaneously learn for classification, localization, and detection of objects. By an efficient sliding-window approach, the network is run across each image location and for multiple scales such that the computations for overlapping regions are done only once. This way, OverFeat is much speedier than DetectorNet. Next, the classification layer of softmax is once again replaced by a regression layer and thereafter it is trained to predict bounding boxes. The classifier and the regressor networks are sequentially trained to generate probable class label, its confidence (from softmax output) and the horizontal/vertical shifts of each bounding-box proposals.

Single-shot multibox detector (SSD) by Liu *et al.* [15] used VGG-16 architecture as a backbone because it performed the best in image classification and transfer learning tasks at that time. It is a single-shot object detection model. In place of the fully-connected layers at the bottom of VGG-16, the authors introduced multiple auxiliary convolutional feature layers. Each auxiliary layer is responsible for decreasing the hidden layer output progressively to different output sizes. This is used to estimate the confidence scores and offset values to varied scales of default boxes possessing disparate aspect ratios. As such, using multi-scale feature maps and convolutional kernels for the task of detection, SSD could proffer high accuracy with highly-improved computation speed. Feature fusion SSD (FSSD) [16] was an improvement on top of SSD to include a lightweight add-on module for the purpose of fusing features. In SSD, features pertaining to various scales could not be fused efficiently because it used feature pyramid detection. This was improved in FSSD where, features from different layers pertaining to separate scales were concatenated together by feature fusion, to generate new feature pyramids thereafter. This gives a rise in accuracy but a drop in the detection speed, when compared to SSD. Attentive SSD (ASSD) [17] uses a scheme of incorporating attention whereby, significant regions on feature maps were preserved and emphasized while doing away with unnecessary regions.

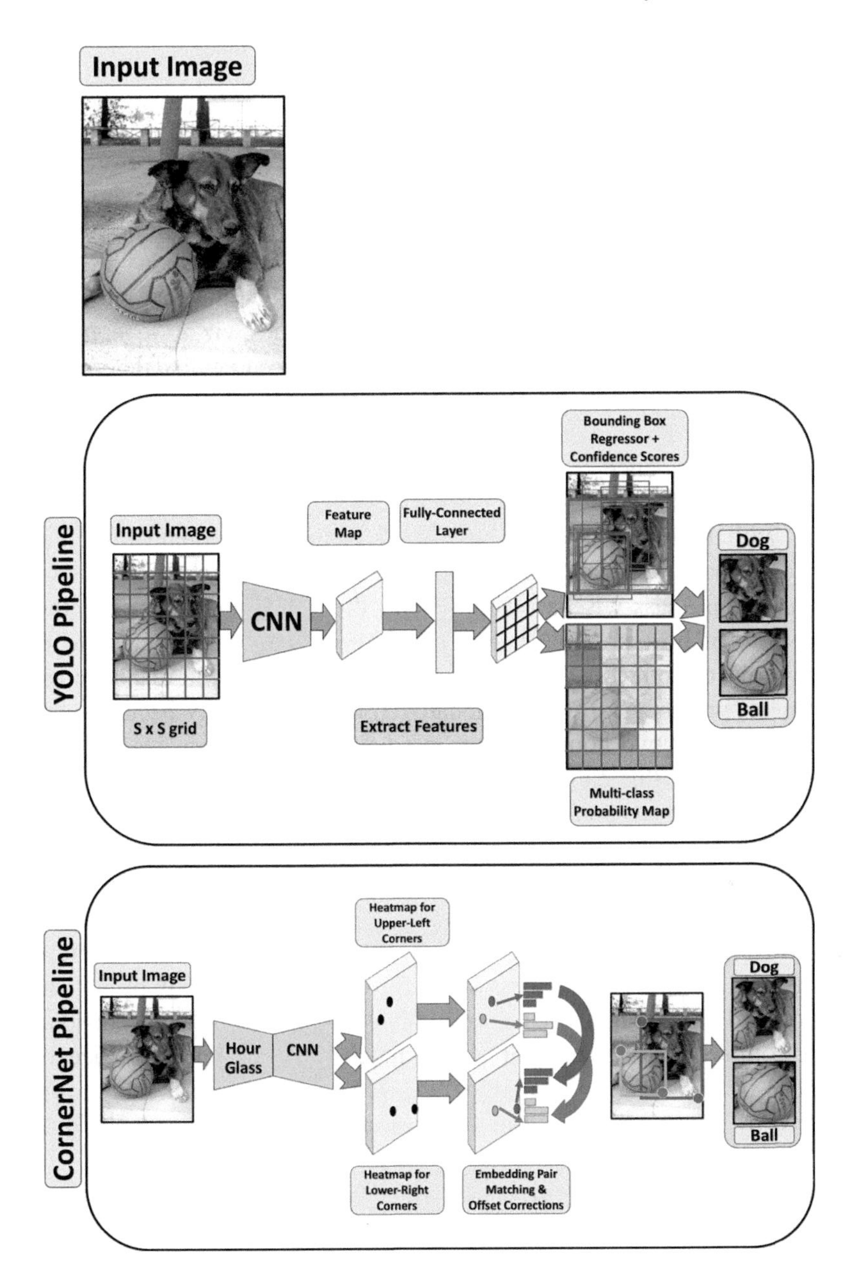

Figure 9.3 Comparison of YOLO and CornerNet pipeline (both monolithic architectures) for an image from COCO [1].

Redmon *et al.* [18] came up with another significant breakthrough in the domain of single-shot 2D object detection. Unlike many earlier approaches, this model entitled *You Only Look Once (YOLO)*, was not a classifier disguised and modified for object detection task. Moreover, this was capable of detecting objects in real-time speed. Its overall framework is shown in as illustrated in Figure 9.3. Here, the input image is first partitioned into a grid each of whose cell is $S \times S$. Each such cell has the responsibility of classifying and localizing an object whose center lies within that particular cell. A bounding-box prediction is defined by five values, viz. x, y, w, h, CF where x, y represents the spatial coordinate of the box center, w, h determines the height and width of that box and CF (objectness score) is a measure of how confident the architecture is about presence of an object within that cell. The confidence score is obtained by intersection-over-union ratio with a ground-truth bounding box and for overlapping proposals, only the proposal with highest CF is preserved to be computationally efficient. Finally, each box proposal is separately classified to get probability over all classes. This class-wise probability and the objectness score together gives a score of how possible it is for an object of particular class to be present within the bounding-box proposal. But, YOLO has a disadvantage that it cannot predict small objects located at close proximity because each cell can only have two bounding box predictions of a single class. Many improvements over this came in the subsequent years. One such approach was YOLOv2 [19] that catered to improving recall and localization in YOLO by changes like introducing batch normalization, increasing input resolution, prediction of multiple objects per grid-cell with anchors, etc. Fast YOLO [20] improved upon YOLOv2 architecture where motion-adaptive inference was employed to attain even faster detection with a more compact architecture. YOLOv3 [21] computes objectness score for each bounding box by logistic regression and uses a relatively larger architecture DarkNet-19 with residual shortcut connections. As such, it shows much improved performance on small close-by objects as compared to YOLO. YOLOv4 [22] is a more faster alternative by incorporating bag-of-freebies and bag-of-freebies techniques during the training phase.

Other popular one-stage object detection frameworks include CornerNet [23] which does not necessitate formulation of good anchor boxes. Using a single-CNN pipeline, they predict separate heatmaps for upper-left and lower-right corners of all object instances that belong to the same class. Further, the CNN is trained to predict an embedding vector with each such corner points, such that corresponding pair of upper-left and lower-right corners for the same objects possess a minimal inter-embedding distance. Such pairs are thereby matched (as illustrated in Figure 9.3) to form a bounding-box encompassing an object. Apart from CNNs, transformers [24] are also now being used in 2D object detection and it has shown promising outcomes. Carion *et al.* [25] formulated the 2D object detection problem by a direct set prediction approach. Similar to CornerNet, this also does not require anchor boxes. First, a CNN architecture generates 2D feature matrix that represents an input image. After flattening, this feature matrix is augmented with positional encoding. This is passed through a encoder–decoder transformer framework that proffers a set of output embeddings corresponding to object queries. A shared feed-forward network predicts the class-label and object bounding box, or determines if the box represents no object inside it. The whole architecture is trained end-to-end with bipartite matching loss.

9.2.2.2 Region proposal-based two-stage architectures for 2D object detection

In region-proposal-based architectures, object detection is done in two stages. The first stage is class-agnostic where multiple region-proposals are generated from an image, without any regard toward the class to which it can belong. Thereafter, as part of the second stage, a ConvNet is used to extract features from these proposals and thereby, classify them into object categories.

The region-based convolutional neural networks (R-CNN), conceptualized by Girshick *et al.* [26], was a major landmark in the history of CNN-based object detection. The method contains three stages. As part of first stage of the R-CNN, a large number of region proposal ($\sim$2000) are generated by selective search on the input image. These proposals are independent of the object class to which they belong. Next, in the second stage, features are extracted from each such region proposal through a CNN, pre-trained on the task of image classification. To ensure that the input region proposals are compatible with the square-shaped fixed input shape of CNN, they are warped to a compact bounding box around. The CNN is fine-tuned on these region proposals for all object classes and one extra class indicating background class (i.e., region proposal contains none of the intended objects). Finally, as part of the third stage, there are separate binary (positive/negative) support vector machines (SVMs) trained independently for each object class. These SVMs take in the feature vector from the last layer of CNN for a region proposal, and scores the proposals. Through non-maximum suppression (NMS), regions with IoU overlap above a certain threshold are discarded and remaining are preserved. Further, the CNN features are also used to train a regression network to rectify and tighten the precise location and size of the bounding box. Although this was a seminal work in itself, one of its major disadvantages was its computation time-complexity. Performing selective search on each input image to generate ∼2000 region proposals and passing each of them through CNN to generate features is a time-consuming process. Further, the three stages are independent of each other and thus, do not share weights among them. This makes the overall object detection process slow.

He *et al.* [27] introduced spatial pyramid pooling in CNN (SPPNet) to detect objects in images. In R-CNN, the last fully-connected layer of CNN necessitated the input image to be of fixed size. To do away with this, the authors added a spatial pyramid pooling layer after the final convolutional layer, such that the input to the fully-connected layer is of fixed dimension. The overall object detection process gets faster, because for every input image, the CNN needs to do a single pass. But, the training is again time-consuming and weights of convolutional layers prior to spatial pyramid pooling layer are not updated during fine-tuning the deep neural network architecture on the region proposals.

Girshick [28] later improved upon R-CNN and proposed fast R-CNN. The idea is similar to R-CNN but here, instead of three separate stages, there is a unified process now with sharing of CNN weights. Unlike R-CNN, here the CNN does not take region proposals as input. Instead, the CNN accepts the raw RGB image as input. The CNN thereafter, through its multiple convolutional and max pooling layers, generates a convolutional feature map. Region proposals are generated from this convolutional feature map and a region-of-interest (RoI) pooling generates a fixed-size feature vector for each such region proposal. A fully-connected layer accepts this feature vector and does two parallel predictions. The first is through a softmax layer corresponding to all object classes and one extra class indicating background. The second parallel prediction is to predict and refine the four corners of the bounding box of region proposal.

Another subsequent improvement by Ren *et al.* [29] is through faster R-CNN. Unlike R-CNN and fast R-CNN, there is no separate selective search step here to generate object proposals. Instead, they use fully convolutional networks to learn generating region proposals. Keeping the proposals same, there is alteration between fine-tuning region proposal and object detection. Thereby, the weights of the CNN are shared between both these tasks and the process is even faster than earlier.

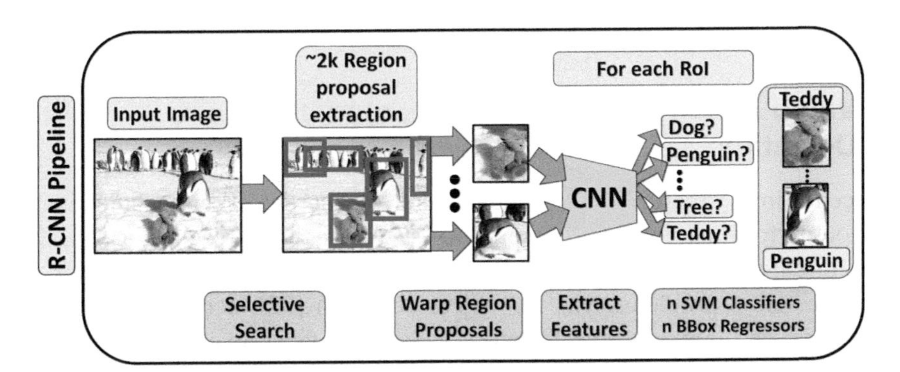

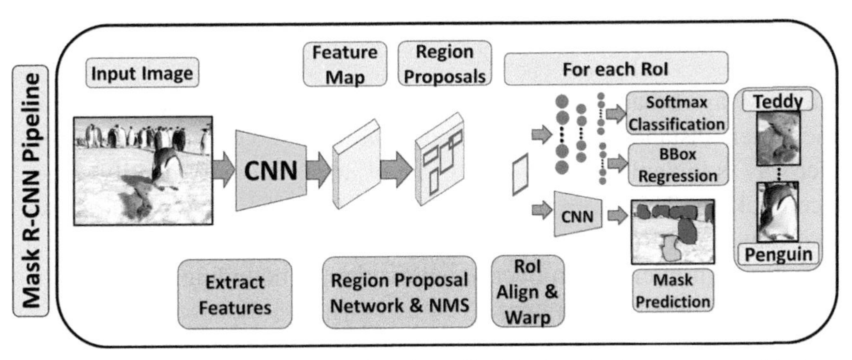

Figure 9.4 Comparison of R-CNN and mask R-CNN pipeline (both two-stage architectures) for an image from COCO [1].

Mask R-CNN was later proposed by He *et al.* [30], as an extension on top of faster R-CNN. We compare its framework with vanilla R-CNN in Figure 9.4. As per the first stage of mask R-CNN, a region proposal network generates region of interests in similar lines to faster R-CNN. In the second stage, the bounding box and object category is predicted for each region of interest generated in the first stage. Along with this, mask R-CNN has an additional parallel pipeline that generates object masks. These masks are similar to performing semantic segmentation [31] on the whole image to generate pixel-wise classification. RoIPool serves extraction of feature map from each region of interest. But this may lead to quantization errors, which are handled by a RoIAlign layer (containing two convolutional layers) that aligns feature maps to the original input image. This in turn, generates pixel-level segmentation of each region of interest along with the original object detection pipeline of faster R-CNN.

9.3 Video Activity Recognition Assisted by Object Detection

In recent times, there is an abundance of object detection techniques, each outperforming its predecessors. As such, we have reached a stage where researchers are moving on to improve performance on even complex computer vision problems, viz. 3D object detection [32], action detection, localization and tracking objects across videos [4], video event recognition and scene understanding [33][34], analysis of facial expressions for smart healthcare systems [35], etc. In this chapter, we focus on the task of event/activity recognition in videos, done with the assistance of frame-wise object detection, which enables inter-frame tracking of objects. This can be done in two ways broadly. The first way is to use per-frame object detection and thereafter, instance identification such as to track the same instances. Another way is to apply multi-frame fusion of information to identify same-instance tracking. This is particularly effective because every object shows minimal movement from one frame to the next. As such, there should be a significantly high intersection-over-union overlap of bounding boxes for the same object, in consecutive frames.

Burić *et al.* [36] employed object detection as a prerequisite for efficient activity recognition in sports videos. Specifically, they consider the team sports of handball and identify salient objects like players, ball, etc. Videos were recorded by GoPro camera in complex, challenging environments of sports hall/complex (for indoor) and playground (for outdoor) that exhibits varied artifacts like cluttering, shadows, low illumination, and jitters. They compared three methods, viz. frame-wise CNN-based object detection like YOLOv2, Mask R-CNN, and moving foreground-object detector like mixture of Gaussians (MoG) [37]. The principle behind MoG method is background subtraction, which is specifically useful in videos where the background is relatively static. By averaging the RGB values of all frames, we get an estimate of background. This background, when subtracted from each frame, proffers the foreground region in the frame. MoG is more robust toward non-static backgrounds, where history of each pixel is modeled through an approximation by K-Means clustering on the Expectation-Maximization algorithm. Results showed that YOLOv2 always predicted bounding box around salient foreground objects, but had difficulty in distinguishing objects camouflaged with its immediate background. MoG was oftentimes erroneous in this task, because apart from real objects, it detected bounding box around background patches due to illumination changes for example, reflections, floodlights etc. Nevertheless, it did not face issues in camouflaging objects as it is driven by motion, not colors. Mask R-CNN was better in detecting smaller objects but similar to YOLOv2, it faced difficulties in detecting the ball (which was not an issue in MoG). However, none of the three methods were confused by shadows and did not end up detecting shadows as real objects.

Ivasic-Kos *et al.* [38] suggested methodologies to track object instances across frames. This can be specifically helpful in video event detection too. They use mask R-CNN to detect players in individual frames of handball videos. Thereafter, this detection and location information is combined with a measure of the players' activity by spatio-temporal interest points. This helps in tracking the same player throughout the entire video.

Gleason *et al.* [39] discussed action classification and spatio-temporal action detection in videos with the help of object detection. First, they perform frame-wise object detection by mask R-CNN. These detection results are aggregated by hierarchical clustering approach whereby object detection bounding boxes are represented by a 3D coordinate comprised of frame number, x-coordinate and y-coordinate of the bounding-box center, and thereby clustered. Thereafter, by temporal jittering, dense action proposals are generated from the clusters proffered by hierarchical clustering. Finally, action classification was performed by training a Temporal Refinement I3D architecture that simultaneously classifies actions and refines the time-bounds for detected actions.

9.4 Recent Datasets on 2D Object Detection

Research in deep learning-based object detection flourished through the introduction of large-scale datasets containing fine-grained annotations of varied objects. In this regard, PASCAL VOC [40] and MS COCO [1] are two datasets that have been used extensively in various object detection research. PASCAL VOC contains ∼11,000 images spread over 20 categories which include person and subcategories of animal (e.g., bird, horse), indoor objects (e.g., boat, train), and vehicles (e.g., sofa, tv). MS COCO is an even larger dataset with about 328,000 images spread over 91 object categories, which include the PASCAL VOC classes and also some rarely occurring categories such as toothbrush, toaster, wine glass, etc.

More datasets are being introduced regularly by the research community, that are even larger in terms of scale and cater to detection of both general objects and specialized objects such as guns, camouflaged objects, etc. We now present some of the datasets for generalized and specialized object detection in images, which have gained attention in recent times:

Dataset for Object deTection in Aerial Images (DOTA) [41]. This is a large-scale dataset intended for the task of detecting objects in aerial images. In its version 2.0, there are 11,268 images covering about 1.8 million object instances. Images are collected mainly from Google Earth, CycloMedia, and other organizations that collect satellite and aerial images. Object categories range across 18 classes involving *transportation* like large or small vehicles, ships, helicopters, *architectures* like bridge, harbors, *ground patches* like tennis courts, swimming pool, helipad, etc.

Objects in Context in Overhead Imagery (xView) [42]. This dataset is for object detection tasks in complex overhead scenes, in the form of images. The aim of such a dataset is to support models that can identify areas hit hard by catastrophes and disasters and thereby can speed-up the response and disaster management. Images are in the form of satellite images taken from WorldView-3 satellites. In total, the images contain about 1 million objects spread over 60 object categories. Objects range from commonly spotted real-life instances like cars, buildings to land-use locations like vehicle lot, construction site, helipad and large objects like bus, trailer, tank car, etc.

Object365 [43]. Object365 is a dataset focused toward supporting detection of a diverse range of objects in the wild. This dataset contains around 30 million object instances spread over 2 million images. The number of object classes is 365, which can be grouped under 11 high-level categories. The high-level categories include *human & accessory* (e.g., person, watch, book), *clothes* (e.g., slippers, mask, hat), *kitchen items* (e.g., plate, refrigerator, knife), *animals* (e.g., crab, yak, dolphin), *food* (e.g., carrot, avocado, coconut), *electronics* (e.g., printer, microphone, compact disk), etc.

Open Images Dataset [44]. The dataset comprises almost 16 million object instances on 1.9 million images. The objects can be grouped into 600 object classes. The objects cover a varied range of real-world entities. Objects include course-grained categories (e.g., animals), fine-grained categories (e.g., shellfish), scenes, events (e.g., birthday, wedding), materials, etc.

Camouflaged Object Detection (COD10k) [45]. This dataset is composed of images collected from photography websites (e.g., Flickr), where the intended objects to be detected are obscured organisms within its surroundings. There are around 10,000 images, with annotated object bounding boxes. The objects can be categorized into 78 classes (out of which 69 classes are of camouflaged objects, 9 classes are of non-camouflaged objects), which can be again grouped into 10 super-classes (out of which 5 super-classes are of camouflaged objects, rest are of non-camouflaged objects). Object examples include dog, fish, mantis, toad, deer, cicada, etc., which are grouped into super-classes like aquatic animals, flying animals, terrestrial animals, amphibians, and others.

Large Vocabulary Instance Segmentation (LVIS) [46]. The dataset contains approximately 2 million object instance annotations across 0.16 million images. The images were mostly obtained from the COCO dataset [1]. Objects are categorized into 1000 categories of regular everyday objects, e.g., bed, tea-cup, peanut, donuts, umbrella, shoulder bag.

Further, we elaborate on some recent datasets for object detection in video clips:

YouTube-BoundingBoxes (YT-BB) [47]. Unlike the aforementioned datasets, this dataset focuses on object detection in videos. It comprises about 5.6 million object instances (annotated in the form of bounding box) across 380,000 trimmed videos of approximately 15–20 seconds duration, which specifically features objects appearing in natural scenes. Videos were sourced from YouTube. Video objects are categorized into 23 object classes, encompassing, person, animals (e.g., dog, elephant), or other common objects (e.g., aeroplane, crowd).

Temporal Hands Guns and Phones Dataset (THGP) [48]. This is also a dataset catering to object detection in videos. In the training partition, there are 50 videos, each of 100 frames. The test partition, on the other hand, contains 48 videos of 20 frames each. The ongoing events in the videos are specific to policing and crimes for example, armed robberies in stores, shooting drills, making calls. As such, the objects like hands, guns, and phones are annotated in the video frames.

Objects Around Krishna (OAK) [49]. This dataset comprises of 80 video clips spanning about 17 hours. Bounding-box annotations are provided for detecting objects continually, in the video frames. Videos are egocentric in nature and were captured by graduate students over a long period of time by a camera called KrishnaCam. Since the videos represent regular outdoor scenes, the objects are effectively natural everyday objects, which are categorized into 105 object classes.

Table 9.1 Performance (mAP) of state-of-the-art 2D object detection methods in images, on some of the widely used benchmark datasets.

Method	CNN Backbone	PASCAL VOC [40]	MS COCO [1]
Monolithic Framework			
SSD513 [15]	ResNet-101	76.8	31.2
DSSD513 [50]	ResNet-101	81.5	33.2
YOLO [18]	∼GoogLeNet	66.4	-
YOLOv2 [19]	DarkNet-19	78.6	21.6
YOLOv3 [21]	DarkNet-53	-	33.0
YOLOv4 [22]	CSPDarkNet-53	-	43.5
CornerNet511 [23]	Hourglass-104	-	42.1
RetinaNet [51]	ResNet-101	-	39.1
EfficientDet-D7 [52]	BiFPN	-	52.2
DETR-DC5 [25]	ResNet-101	-	44.9
Swin Transformer [53]	HTC++	-	58.7
Two-Stage Framework			
SPP-Net [27]	ZF-5	60.9	-
R-CNN [26]	AlexNet	58.5	-
Fast R-CNN [28]	VGG-16	70.0	19.7
Faster R-CNN [29]	VGG-16	73.2	21.9
FPN [54]	ResNet-101	-	36.2
Mask R-CNN [30]	ResNeXt-101	-	39.8
Cascade R-CNN [55]	ResNeXt-101	-	45.8
DetectoRS [56]	ResNeXt-101	-	55.7

9.5 Performance of Object Detection Techniques

In Table 9.1, we compare the performance of some of the state-of-the-art object detection techniques on the PASCAL VOC [40] and MS COCO [1] datasets. As evident from Table 9.1, PASCAL VOC was mostly used for experimentation in the earlier research papers. Its predominance was later replaced by MS COCO dataset, which contains more object annotations and is a superset of all the PASCAL VOC object categories. For PASCAL VOC, the metric used for comparison is interpolated average precision. The threshold of intersection-over-union (IoU) for predicted and ground-truth bounding boxes is 0.5. In case of MS COCO, the metric used for comparison is mean Average Precision (mAP) performance, which is computed by averaging performance over 10 IoUs in the range [0.5, 0.95] at intervals of 0.05.

9.6 Conclusion and Way Forward

In this chapter, we elaborate upon some of the recent state-of-the-art object detection techniques in use. Monolithic architecture exhibit a single unified ConvNet and/or Transformer pipeline for bounding-box regression and object classification. Two-stage architectures have a separate region-proposal step whose extracted features are used to predict object category and bounding-box coordinates. We compare and contrast some of the significant monolithic and two-stage architecture-based object detection techniques. We also discuss the usefulness of using frame-wise object detection in videos for the task of activity recognition and video event classification. Finally, the chapter details some of the recent large-scale datasets for 2D object detection in images and videos, useful for training object detection frameworks.

In future, trends of object detection would most likely include more few-shot or zero-shot learning approaches to do away with the need for large-scale annotated datasets, catering to both commonly- and rarely-occurring object categories. Researchers are also moving toward more explainable object detection approaches to avoid inadvertent false negatives in critical emergency deployments like 24×7 surveillance, rescue operation etc. For applications in autonomous vehicles and indoor scene understanding, 3D object detection in point clouds is another emerging research area, that is used to recognize and localize objects in 3D environments.

References

[1] Tsung-Yi Lin, Michael Maire, Serge Belongie, James Hays, Pietro Perona, Deva Ramanan, Piotr Dollár, and C Lawrence Zitnick. Microsoft COCO: Common Objects in Context. In *European Conference on Computer Vision*, pages 740–755. Springer, 2014. Available: https://cocodataset.org/, last accessed: Dec, 2021.

[2] AJ Piergiovanni and Michael S Ryoo. Fine-grained Activity Recognition in Baseball Videos. In *Proc. IEEE Conference on Computer Vision and Pattern Recognition workshops*, pages 1740–1748, 2018. Available: https://github.com/piergiaj/mlb-youtube/, last accessed: Dec, 2021.

[3] Swarnabja Bhaumik, Prithwish Jana, and Partha Pratim Mohanta. Event and Activity Recognition in Video Surveillance for Cyber-Physical Systems. In *Emergence of Cyber Physical System and IoT in Smart Automation and Robotics*, pages 51–68. Springer, 2021.

[4] Prithwish Jana, Swarnabja Bhaumik, and Partha Pratim Mohanta. Unsupervised Action Localization Crop in Video Retargeting for 3D ConvNets. In *2021 IEEE Region 10 Conference (TENCON)*. IEEE, 2021.

[5] Prithwish Jana, Swarnabja Bhaumik, and Partha Pratim Mohanta. Key-frame based Event Recognition in Unconstrained Videos using Temporal Features. In *Proc. IEEE Region 10 Symposium (TENSYMP)*, pages 349–354. IEEE, 2019.

[6] Martin A Fischler and Robert A Elschlager. The Representation and Matching of Pictorial Structures. *IEEE Trans. on Computers*, 100(1):67–92, 1973.

[7] Michael C Burl, Markus Weber, and Pietro Perona. A Probabilistic Approach to Object Recognition using Local Photometry and Global Geometry. In *European Conference on Computer Vision*, pages 628–641. Springer, 1998.

[8] Pham Ich Quy and Martin Polasek. Using Thresholding Techniques for Object Detection in Infrared Images. In *Proc. 16th International Conference on Mechatronics-Mechatronika 2014*, pages 530–537. IEEE, 2014.

[9] Rui Miao, Hongxu Jiang, Fangzheng Tian, Xiaobin Li, Yonghua Zhang, and Dong Dong. Real-Time Ship Detection from Infrared Images Through Multi-feature Fusion. In *2021 IEEE Int'l Conf. on Parallel & Distributed Processing with Applications, Big Data &*

Cloud Computing, Sustainable Computing & Communications, Social Computing & Networking, pages 533–538. IEEE, 2021.

[10] Prithwish Jana, Soulib Ghosh, Ram Sarkar, and Mita Nasipuri. A Fuzzy C-means based Approach Towards Efficient Document Image Binarization. In *2017 Ninth International Conference on Advances in Pattern Recognition (ICAPR)*, pages 332–337. IEEE, 2017.

[11] Ting L Luo, Marisa C Eisenberg, Michael AL Hayashi, Carlos Gonzalez-Cabezas, Betsy Foxman, Carl F Marrs, and Alexander H Rickard. A Sensitive Thresholding Method for Confocal Laser Scanning Microscope Image Stacks of Microbial Biofilms. *Scientific Reports*, 8(1):1–14, 2018.

[12] Christian Szegedy, Alexander Toshev, and Dumitru Erhan. Deep Neural Networks for Object Detection. 2013.

[13] Alex Krizhevsky, Ilya Sutskever, and Geoffrey E Hinton. ImageNet Classification with Deep Convolutional Neural Networks. In *Advances in Neural Information Processing Systems*, volume 25, pages 1097–1105, 2012.

[14] Pierre Sermanet, David Eigen, Xiang Zhang, Michaël Mathieu, Rob Fergus, and Yann LeCun. OverFeat: Integrated Recognition, Localization and Detection using Convolutional Networks. In *International Conference on Learning Representations (ICLR)*, 2014.

[15] Wei Liu, Dragomir Anguelov, Dumitru Erhan, Christian Szegedy, Scott Reed, Cheng-Yang Fu, and Alexander C Berg. SSD: Single Shot Multibox Detector. In *European Conference on Computer Vision*, pages 21–37. Springer, 2016.

[16] Zuoxin Li and Fuqiang Zhou. FSSD: Feature Fusion Single Shot Multibox Detector. *arXiv preprint arXiv:1712.00960*, 2017.

[17] Jingru Yi, Pengxiang Wu, and Dimitris N Metaxas. ASSD: Attentive Single Shot Multibox Detector. *Computer Vision and Image Understanding*, 189:102827, 2019.

[18] Joseph Redmon, Santosh Divvala, Ross Girshick, and Ali Farhadi. You Only Look Once: Unified, Real-time Object Detection. In *Proc. IEEE Conference on Computer Vision and Pattern Recognition*, pages 779–788, 2016.

[19] Joseph Redmon and Ali Farhadi. YOLO9000: Better, Faster, Stronger. In *Proc. IEEE Conference on Computer Vision and Pattern Recognition*, pages 7263–7271, 2017.

[20] Mohammad Javad Shafiee, Brendan Chywl, Francis Li, and Alexander Wong. Fast YOLO: A Fast You Only Look Once System for Real-time Embedded Object Detection in Video. *arXiv preprint arXiv:1709.05943*, 2017.

[21] Ali Farhadi and Joseph Redmon. YOLOv3: An Incremental Improvement. In *Computer Vision and Pattern Recognition*, pages 1804–2767. Springer Berlin/Heidelberg, Germany, 2018.

[22] Alexey Bochkovskiy, Chien-Yao Wang, and Hong-Yuan Mark Liao. YOLOv4: Optimal Speed and Accuracy of Object Detection. *arXiv preprint arXiv:2004.10934*, 2020.

[23] Hei Law and Jia Deng. CornerNet: Detecting Objects as Paired Keypoints. In *Proc. European Conference on Computer Vision (ECCV)*, pages 734–750, 2018.

[24] Thomas Wolf, Lysandre Debut, Victor Sanh, Julien Chaumond, Clement Delangue, Anthony Moi, Pierric Cistac, *et al.* Transformers: State-of-the-art Natural Language Processing. In *Proc. 2020 Conference on Empirical Methods in Natural Language Processing: System Demonstrations*, pages 38–45, 2020.

[25] Nicolas Carion, Francisco Massa, Gabriel Synnaeve, Nicolas Usunier, Alexander Kirillov, and Sergey Zagoruyko. End-to-End Object Detection with Transformers. In *European Conference on Computer Vision*, pages 213–229. Springer, 2020.

[26] Ross Girshick, Jeff Donahue, Trevor Darrell, and Jitendra Malik. Rich Feature Hierarchies for Accurate Object Detection and Semantic Segmentation. In *Proc. IEEE Conference on Computer Vision and Pattern Recognition*, pages 580–587, 2014.

[27] Kaiming He, Xiangyu Zhang, Shaoqing Ren, and Jian Sun. Spatial Pyramid Pooling in Deep Convolutional Networks for Visual Recognition. *IEEE Trans. on Pattern Analysis and Machine Intelligence*, 37(9):1904–1916, 2015.

[28] Ross Girshick. Fast R-CNN. In *Proc. IEEE International Conference on Computer Vision*, pages 1440–1448, 2015.

[29] Shaoqing Ren, Kaiming He, Ross Girshick, and Jian Sun. Faster R-CNN: Towards Real-time Object Detection with Region Proposal Networks. *IEEE Trans. on Pattern Analysis and Machine Intelligence*, 39(6):1137–1149, 2016.

[30] Kaiming He, Georgia Gkioxari, Piotr Dollár, and Ross Girshick. Mask R-CNN. In *Proc. IEEE International Conference on Computer Vision*, pages 2961–2969, 2017.

[31] Aritra Mukherjee, Prithwish Jana, Sayak Chakraborty, and Sanjoy Kumar Saha. Two Stage Semantic Segmentation by SEEDS and Fork Net. In *2020 IEEE Calcutta Conference (CALCON)*, pages 283–287. IEEE, 2020.

[32] Ishan Misra, Rohit Girdhar, and Armand Joulin. An End-to-End Transformer Model for 3D Object Detection. In *Proc. IEEE/CVF International Conference on Computer Vision*, pages 2906–2917, 2021.

[33] Saiyed Umer, Mrinmoy Ghorai, and Partha Pratim Mohanta. Event Recognition in Unconstrained Video using Multi-scale Deep Spatial Features. In *2017 Ninth International Conference on Advances in Pattern Recognition (ICAPR)*. IEEE, 2017.

[34] Prithwish Jana, Swarnabja Bhaumik, and Partha Pratim Mohanta. A Multi-Tier Fusion Strategy for Event Classification in Unconstrained Videos. In *International Conference on Pattern Recognition and Machine Intelligence (PReMI)*, pages 515–524. Springer, 2019.

[35] Anay Ghosh, Saiyed Umer, Muhammad Khurram Khan, Ranjeet Kumar Rout, and Bibhas Chandra Dhara. Smart Sentiment Analysis System for Pain Detection using Cutting Edge Techniques in a Smart Healthcare Framework. *Cluster Computing*, 2022.

[36] Matija Burić, Miran Pobar, and Marina Ivašić-Kos. Object Detection in Sports Videos. In *2018 41st International Convention on Information and Communication Technology, Electronics and Microelectronics (MIPRO)*, pages 1034–1039. IEEE, 2018.

[37] Chris Stauffer and W Eric L Grimson. Adaptive Background Mixture Models for Real-time Tracking. In *Proceedings. 1999 IEEE Computer Society Conference on Computer Vision and Pattern Recognition*, volume 2, pages 246–252. IEEE, 1999.

[38] Marina Ivasic-Kos and Miran Pobar. Building a Labeled Dataset for Recognition of Handball Actions using Mask R-CNN and STIPS. In *2018 7th European Workshop on Visual Information Processing (EUVIP)*. IEEE, 2018.

[39] Joshua Gleason, Rajeev Ranjan, Steven Schwarcz, Carlos Castillo, Jun-Cheng Chen, and Rama Chellappa. A Proposal-based Solution to Spatio-Temporal Action Detection in Untrimmed Videos. In *2019 IEEE Winter Conference on Applications of Computer Vision (WACV)*, pages 141–150. IEEE, 2019.

[40] Mark Everingham, SM Eslami, Luc Van Gool, Christopher KI Williams, John Winn, and Andrew Zisserman. The PASCAL Visual Object Classes

Challenge: A Retrospective. *International Journal of Computer Vision*, 111(1):98–136, 2015.

[41] Jian Ding, Nan Xue, Gui-Song Xia, Xiang Bai, Wen Yang, Micheal Ying Yang, Serge Belongie, Jiebo Luo, Mihai Datcu, Marcello Pelillo, and Liangpei Zhang. Object Detection in Aerial Images: A Large-Scale Benchmark and Challenges, 2021. Available: https://captain-whu.github.io/DOTA/, last accessed: Dec, 2021.

[42] Darius Lam, Richard Kuzma, Kevin McGee, Samuel Dooley, Michael Laielli, Matthew Klaric, Yaroslav Bulatov, and Brendan McCord. xview: Objects in context in overhead imagery. *arXiv preprint arXiv:1802.07856*, 2018. Available: http://xviewdataset.org/, last accessed: Dec, 2021.

[43] Shuai Shao, Zeming Li, Tianyuan Zhang, Chao Peng, Gang Yu, Xiangyu Zhang, Jing Li, and Jian Sun. Objects365: A Large-scale, High-quality Dataset for Object Detection. In *Proc. IEEE/CVF International Conf. on Computer Vision*, pages 8430–8439, 2019. Available: http://www.objects365.org/, last accessed: Dec, 2021.

[44] Alina Kuznetsova, Hassan Rom, Neil Alldrin, Jasper Uijlings, Ivan Krasin, et al. The Open Images Dataset v4. *International Journal of Computer Vision*, 128(7):1956–1981, 2020. Available: https://storage.googleapis.com/openimages/web/2018-05-17-rotation-information.html, last accessed: Dec, 2021.

[45] Deng-Ping Fan, Ge-Peng Ji, Guolei Sun, Ming-Ming Cheng, Jianbing Shen, and Ling Shao. Camouflaged Object Detection. In *Proc. IEEE/CVF Conference on Computer Vision and Pattern Recognition*, pages 2777–2787, 2020. Available: https://github.com/DengPingFan/SINet/, last accessed: Dec, 2021.

[46] Agrim Gupta, Piotr Dollar, and Ross Girshick. LVIS: A Dataset for Large Vocabulary Instance Segmentation. In *Proc. IEEE/CVF Conference on Computer Vision and Pattern Recognition*, pages 5356–5364, 2019. Available: https://www.lvisdataset.org/, last accessed: Dec, 2021.

[47] Esteban Real, Jonathon Shlens, Stefano Mazzocchi, Xin Pan, and Vincent Vanhoucke. Youtube-BoundingBoxes: A Large High-Precision Human-Annotated Data Set for Object Detection in Video. In *Proc. IEEE Conference on Computer Vision and Pattern Recognition*, pages 5296–5305, 2017. Available: https://research.google.com/youtube-bb/, last accessed: Dec, 2021.

[48] Mario Alberto Duran-Vega, Miguel Gonzalez-Mendoza, Leonardo Chang-Fernandez, and Cuauhtemoc Daniel Suarez-Ramirez. TYOLOv5: A Temporal Yolov5 Detector Based on Quasi-Recurrent Neural Networks for Real-Time Handgun Detection in Video. *arXiv preprint arXiv:2111.08867*, 2021.

[49] Jianren Wang, Xin Wang, Yue Shang-Guan, and Abhinav Gupta. Wanderlust: Online Continual Object Detection in the Real World. In *Proc. IEEE/CVF International Conference on Computer Vision*, pages 10829–10838, 2021. Available: https://oakdata.github.io/, last accessed: Dec, 2021.

[50] Cheng-Yang Fu, Wei Liu, Ananth Ranga, Ambrish Tyagi, and Alexander C Berg. DSSD: Deconvolutional Single Shot Detector. *arXiv preprint arXiv:1701.06659*, 2017.

[51] Tsung-Yi Lin, Priya Goyal, Ross Girshick, Kaiming He, and Piotr Dollár. Focal Loss for Dense Object Detection. In *Proc. IEEE International Conference on Computer Vision*, pages 2980–2988, 2017.

[52] Mingxing Tan, Ruoming Pang, and Quoc V Le. EfficientDet: Scalable and Efficient Object Detection. In *Proc. IEEE/CVF Conf. on Computer Vision and Pattern Recognition*, pages 10781–90, 2020.

[53] Ze Liu, Yutong Lin, Yue Cao, Han Hu, Yixuan Wei, Zheng Zhang, Stephen Lin, and Baining Guo. Swin Transformer: Hierarchical Vision Transformer using Shifted Windows. In *Proc. IEEE/CVF International Conference on Computer Vision*, pages 10012–10022, 2021.

[54] Tsung-Yi Lin, Piotr Dollár, Ross Girshick, Kaiming He, Bharath Hariharan, and Serge Belongie. Feature Pyramid Networks for Object Detection. In *Proc. IEEE Conference on Computer Vision and Pattern Recognition*, pages 2117–2125, 2017.

[55] Zhaowei Cai and Nuno Vasconcelos. Cascade R-CNN: High Quality Object Detection and Instance Segmentation. *IEEE Trans. on Pattern Analysis and Machine Intelligence*, 43(5):1483–1498, 2019.

[56] Siyuan Qiao, Liang-Chieh Chen, and Alan Yuille. DetectoRS: Detecting Objects with Recursive Feature Pyramid and Switchable Atrous Convolution. In *Proc. IEEE/CVF Conference on Computer Vision and Pattern Recognition*, pages 10213–10224, 2021.

10

Survey on Vehicle Detection, Identification and Count using CNN-based YOLO Architecture and Related Applications

Gaurish Garg[1], Shailendra Tiwari[1], Shivendra Shivani[1], and Avleen Malhi[2]

[1]Department of Computer Science & Engineering, Thapar Institute of Engineering and Technology, India
[2] Data science and AI, Bournemouth university, UK
E-mail: shailendra@thapar.edu; amalhi@bournemouth.ac.uk

Abstract

This chapter presents the detailed analysis of detection, identification, and count of vehicles in an image, using convolutional neural networks CNN)-based YOLO (You Only Look Once) architecture. The proposed approach is applied in civil fields like automated toll revenue calculation and the calculation of expected strength of the road. The vehicles were detected, identified, and counted in real time based on the proposed architecture. The identification of any vehicle is often accomplished through sensing element gear. Examples of these sensing element gears include inductive circle finder, infrared symbol, and measuring device indicator. However, to chop the price, we tend to use YOLO architecture trained on the COCO (Common Objects in Context) dataset, and it's capable of detecting Eighty Common Objects in context.

Keywords: YOLO (You only look once) architecture, convolutional neural network (CNN), COCO dataset, machine learning.

10.1 Introduction

This chapter discusses the methodology behind the YOLOv3 architecture for identifying the vehicles and their real-time applications. Identifying any vehicle is often accomplished through sensing element gear like inductive circle finder, infrared symbol, and measuring device indicator. However, these are costlier strategies. To chop the price, we tend to use the YOLO architecture design trained on the palm (Common Objects in Context) dataset, and it's capable of detecting 80 common objects. It is most commonly used with cvlib's detect_common_objects() function. YOLO suggests that it runs in a single iteration [1]. In the present context, some deep learning methods also accelerate the object detection and recognition with high accuracy [2][3].

10.2 Literature Review

YOLOv3 took sensible ideas from others and trained a new classifier network higher than others. YOLOv3 runs considerably quicker than different detection strategies with comparable performance.

The YOLOv3 design is trained on the palm (Common Objects in Context) dataset which is capable of detecting 80 everyday objects.

In the past, YOLO struggled with little objects. However, currently, we tend to see a reversal in this trend. With the new multi-scale predictions, we tend to see YOLOv3 has comparatively high APS performance [4]. It is good at detecting small-sized objects, but its performance is relatively worse on very small, medium, and bigger-sized objects. When we plot accuracy vs speed on the AP50 metric, we see YOLOv3 has vital edges over different detection systems [5]. Namely, it's quicker and higher [6][7].

YOLOv4 is better than YOLOv3 as it can detect objects in real-time and thus can be an efficient algorithm considering the applications in car tracking and speed estimation. The performance of YOLOv4 exceeds the performance of YOLOv3 by 12%. Not only this, but YOLOv4 also includes two-stage object detection and batch normalisation..

10.3 Methodology

The vehicle detection, identification and counting software uses the YOLOv3 architecture, which is trained on the COCO (Common Objects in Context) dataset, and it is capable of detecting 80 everyday objects [8]. YOLO means that it detects objects in a single iteration. It does not iterate through pixels.

The basic YOLO architecture divides the image into $S \times S$ grid cells and runs through all the grid cells simultaneously [5].

The YOLOv3 version is a variant of darknet. Darknet is a framework used to train neural networks. There are initially 53 layers in it. Another 53 layers are used for the detection task [9]. Therefore, a total of 106 layers are used in the convolutional architecture of YOLOv3 [9]. Within the convolutional layers, the algorithm applies 1×1 kernels on the feature maps of three different sizes at three different places in the network [6][10]. These predictions are calculated at three scales by downsampling the image dimensions in moves of 32, 16 and 8. Downsampling is the decrease in spatial resolution while retaining the same image representation, to decrease the data size.

YOLOv4 architecture focuses more on the speed of the neural network for production systems. In parallel computing, YOLOv4 focuses on optimisation [11]. Different neural network types are used in YOLOv4 though the basic architecture is the same in all YOLO versions. In YOLOv4, for the graphical processing unit, convolutional layers CSPResNeXt50 and CSPDarknet53 are used in small groups of 1 to 8 [11]. The YOLOv4 is an extension of YOLOv3 as it consists of YOLOv3, which forms the head, CSPDarknet53, which includes the backbone and SPP (spatial pyramid pooling) and PAN (pan aggregation network) form the neck of the YOLOv4 architecture [11].

YOLOv5 is officially out, but no significant research has been done on it.

Despite the different variants of YOLO, the basic mathematical model remains the same. The first step of the YOLO algorithm includes dividing the input image into grid cells. The image is divided into $S \times S$ grid cells [5]. B number of bounding boxes and a class score C is predicted within each grid cell. The prediction in each bounding box is a tuple of 4 parameters as shown in Figure 10.3.

- The center x of the bounding box
- The center y of the bounding box
- The width w of the bounding box
- The height h of the bounding box

In addition to this, it also predicts a confidence score c of each bounding box. The confidence score is used to determine the surety of the presence of an object within each grid cell. A higher confidence score for a bounding box indicates that the probability of the cell containing an object is higher.

For each image, the YOLO gives an output vector of $S \times S \times (5B+C)$ numbers, and it is the tensor shape of the final convolutional layer. In

Figure 10.1 The image is divided into $S\times S$ grid cells. In this image, we have 7×7 grid cells.

Figure 10.2 Bounding box is predicted within each grid cell.

$S\times S\times(5B+C)$, $S\times S$ corresponds to the number of grid cells, B is the number of bounding boxes, and C is the class score corresponding to the grid cell as shown in Figure 10.1. The bounding box is predicted within each grid cell as shown in Figure 10.2.

Irrespective of the number of bounding boxes within a grid cell, the YOLO algorithm determines only one set of class scores for all the bounding

Figure 10.3 Parameters in bounding box.

boxes in that cell [12]. For an object to be detected, the cell where its center lies will be responsible for the detection of that object. If the cell contains an object, it predicts a probability P_i of the object belonging to every class C_i and $i = 1, 2, 3, ...C$ where P_i is the probability of the object belonging to class C_i given that the cell contains an object as shown in eqn (10.1). Deep within the convolutional neural networks, these predictions were calculated at three different scales by down sampling the image dimensions in moves of 32, 16, and 8.

$$P_i = P\left(\frac{\text{object belongs to class } C_i}{\text{the cell contains an object}}\right). \tag{10.1}$$

While working with COCO dataset, for any grid cell, when YOLO is used with the everyday objects vector, it will generate 80 conditional class probabilities as the YOLO model is trained on the COCO dataset, which is capable of detecting 80 everyday objects. The class score does NOT depend on the number of bounding boxes within a grid cell [12]. For example, if the center of a "bus" lies within a grid cell, that cell will be responsible for detecting and identifying that "bus." For this grid cell, it will predict class scores for all the vehicles like cars, buses, trucks, bicycles etc. and the class probability or the class score for the cell containing a "bus" would be the highest as indicated in Figure 10.4.

It will initially detect all 80 types of everyday objects, but we are concerned with the detection of vehicles only. So, we filter the results and keep the final bounding boxes that contain a vehicle. Alternatively, a new model could have been trained to detect only vehicles. For the efficiency of the YOLO algorithm, it needs to be error-free and thus, this algorithm

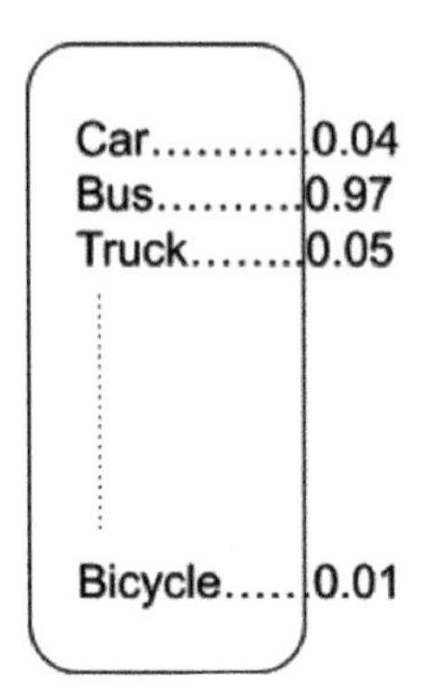

Figure 10.4 Class probabilities vector.

is optimised by the loss function. The loss function works as a localiser. It consists of mean squared error losses [13]. The loss function can be broken into three parts that include:

- The term corresponding to error in finding bounding box coordinates, as expressed in eqn (10.2)
- the term corresponding to error in prediction of score in bounding box as expressed in eqn (10.3)
- the term corresponding to error in prediction of class scores as expressed in eqn (10.4)

$$\lambda_{\text{coord}} \sum_{i=0}^{S^2} \sum_{j=0}^{B} I_{ij}^{\text{obj}} \left[(x_i - x_i)^2 + (y_i - y_i)^2 \right] \tag{10.2}$$

$$\lambda_{\text{coord}} \sum_{i=0}^{S^2} \sum_{j=0}^{B} I_{ij}^{\text{obj}} \left[(\sqrt{w_i} - \sqrt{w_i})^2 + \left(\sqrt{h_i} - \sqrt{h_i} \right)^2 \right] \tag{10.3}$$

$$\begin{aligned} \sum_{i=0}^{S^2} \sum_{j=0}^{B} I_i^{\text{obj}} \Bigg[(c_i - c_i)^2 + \lambda_{noobj} \sum_{i=0}^{S^2} \sum_{j=0}^{B} I_i^{\text{noobj}} \Big[(c_i - c_i)^2 \\ + \sum_{i=0}^{s^2} I_{ij}^{\text{noobj}} \sum_{c \in \text{classes}} \left[\left(p_i(C) - p_i^{\hat{n}}(C) \right)^2 \right] \Big] \Bigg], \end{aligned} \tag{10.4}$$

$$Loss_{\text{yolo}} = (10.2) + (10.3) + (10.4). \tag{10.5}$$

The loss function of YOLO architecture is the summation of eqn (10.2–10.4) as expressed in eqn (10.5). The eqn (10.2) and (10.3) correspond to the bounding box coordinates whereas eqn (10.4) refers to confidence and classification. The members ${I_{ij}}^{\text{obj}}$ and ${I_{ij}}^{\text{noobj}}$ members are used for modulating the loss based on the presence of an object in a particular cell ij [13].

Only the boxes that contain an object within themselves are penalised by the loss function for classification errors if any. Empty boxes are not penalised. The reason for this is that the confidence predictions rely on the differential weights from boxes containing objects and empty boxes during training.

This is all that is beneath the cvlib library's detect_common_objects() function, which returns three parameters, namely bounding box coordinates, corresponding labels, and the confidence scores for the detected objects in the image [13].

10.4 Experimental Results and Applications

On running the algorithm on sample images, we got the results as illustrated in Figures 10.5 through 10.8. The software accurately detected and identified the standard vehicles in the image if the vehicles were not very close.

From the vehicle detection, identification and counting software, it can be observed that it can successfully detect the vehicles and categorise them into cars, buses, trucks, motorcycles or bicycles. It also counts the number of vehicles of a particular type. Also, it can detect and identify the railroad (trains), waterways (boats) and airways (airplane) vehicles as demonstrated in Figures 10.9 and 10.10.

Figure 10.5 Detection of vehicles.

```
Number of cars in the image is 28
Number of bus in the image is 7
Number of truck in the image is 0
Number of bicycle in the image is 0
Number of motorcycle in the image is 0
Number of trains in the image is 0
Number of boats in the image is 0
Number of airplane in the image is 0
```

Figure 10.6 Count of vehicles.

Figure 10.7 Detection of vehicles.

```
Number of cars in the image is 42
Number of bus in the image is 2
Number of truck in the image is 3
Number of bicycle in the image is 0
Number of motorcycle in the image is 0
Number of trains in the image is 0
Number of boats in the image is 0
Number of airplane in the image is 0
```

Figure 10.8 Count of vehicles.

One of the limitations of YOLO is that when there is significantly less space between the vehicles, it becomes difficult for the algorithm to detect all vehicles. The algorithm cannot detect all vehicles when the objects are very close because the YOLO architecture works by dividing the image into an $S \times S$ grid where each cell in the grid predicts only a single object. This is because when multiple close objects are in a single cell, there are high

Figure 10.9 Detection of train.

Figure 10.10 Detection of boat.

chances that some objects won't be detected, thus resulting in inaccurate counting of the objects (vehicles).

Another limitation of this algorithm is that it can detect only cars, buses, trucks, bicycles, motorcycles, trains, boats, and aeroplanes. It cannot uniquely identify vans, jeeps, or special vehicles like auto-rickshaws, ambulances, fire brigades, trams, helicopters, submarines, etc. Ambulances, fire brigades, auto-rickshaws, jeeps, vans are detected but are not uniquely identifiable, i.e., they are identified as trucks. Similarly, trams are detected but identified as trains or buses or trucks. This is because it uses a pre-trained model that does not include special vehicles. Sometimes, this algorithm may detect vans as trucks or cars. If the algorithm used the model trained on the dataset that also contains the special vehicles, then the software would have been accurately able to identify the special vehicles.

Despite the limitations, this vehicle detection, identification and counting algorithm uses various artificial intelligence and machine learning applications. The best example of this is self-driving cars, where the camera mounted on the self-driving car can analyse the road traffic. Thus the self-driving car can make appropriate decisions based on the results processed from the camera input.

Another application includes the detection of vehicles in the wrong lane. It can differentiate between vehicles and detect a truck or bus or motorcycle or bicycle in a car-only lane and alert the traffic control room.

This vehicle detection, identification and counting software can be combined with other computer vision algorithms to implement real-life applications. One of the applications includes camera-based toll tax calculation. This can be achieved by combining this vehicle detection, identification and counting software with the license plate detection software, wherein a vehicle (like a car, bus, truck, motorcycle etc.) and the corresponding license plate number will be identified by the camera at the toll plaza. Then the corresponding toll rate for that vehicle class (car, bus, truck etc.) will be fetched, which will be charged against that detected licence plate number. For example, the vehicle detection software identifies a car at the toll plaza and the licence detection software identifies the licence plate number, say "MH02A0101." Then, the toll rate for the car will be fetched and will be charged against the detected licence plate number [14].

One of the applications includes calculation of toll revenue generated. After vehicle detection, identification and counting of the vehicles, the toll revenue from the vehicles can be calculated as demonstrated in the following Figures 10.11 through 10.13.

For example, in this highway tolling demonstration, if the hypothetical toll is 80 for cars, 100 for buses and 150 for trucks and zero for other categories of road vehicles. For example, consider the number of cars in the input image, which is 28 as obtained from the algorithm as demonstrated in Figure 10.12. So, the amount of toll that can be generated from the number of cars in the input image is $80 \times 28 = 2240$, which is demonstrated in Figure 10.13.

Based on the output images and vehicle count, it can be identified how many cars, buses, trucks or two-wheelers pass a particular stretch of road and using this data, it can be estimated how much weight the road has to bear. This data can be used to estimate the construction material for the widening

Figure 10.11 Detection of vehicles.

```
Number of cars in the image is 28
Number of bus in the image is 7
Number of truck in the image is 0
Number of bicycle in the image is 0
Number of motorcycle in the image is 0
Number of trains in the image is 0
Number of boats in the image is 0
Number of airplane in the image is 0
```

Figure 10.12 Count of vehicles.

```
Suppose we need to calculate total toll revenue
 for different classes of vehicles
Let's say the toll for
car is $80
bus is $100
truck is $150
for bicycle and motorcyle is $0
For image 1
The revenue generated from cars in dollars is
2240
The revenue generated from buses in dollars is
700
The revenue generated from trucks in dollars is
0
The revenue generated from bicycles in dollars is
0
The revenue generated from motorcycles in dollars is
0
The total revenue generated is $2940
```

Figure 10.13 Toll of vehicles.

of roads from road strength. Suppose the average weight of the vehicles is listed in the following Table 10.1.

Table 10.1 Weight of vehicles.

Vehicle type	Average weight at zero capacity (in kg)	Average weight at full capacity (in kg)
Car	1302	1612
Bus	12020.198	15244.198
Truck	11793.402	16629.402
Two-wheeler	181	305

Another application of vehicle detection is the detection of the illegal presence of vehicles in no entry and pedestrian-only zones, In such zones, if a vehicle is detected, the system can alert the traffic control room, and it can be combined with license plate detection software to issue a ticket against the violating vehicle.

For the image in Figure 10.11 and its count in Figure 10.12, and based on average weight in Table 10.1, the total mass on the road when vehicles are at full capacity can be estimated to be $1612 \times 28 + 15244.198 \times 7 = 151,845.386$ kg. Assuming the acceleration due to gravity to be $9.8\ \mathrm{ms}^{-2}$, the total weight on the road can be estimated to be 1488084.78 N (where N refers to Newton). Since the road in Figure 10.11 has six lanes and average width of a six-laned road is 3.75×6 is 22.5 meters $(= b)$, thus tensile strength (or breakdown strength or the strength at which the road fails) can be calculated as $\sigma = \frac{w}{A}$ where σ refers to tensile strength, *W* refers to total weight on the road and *A* refers to the cross-sectional area of the road. For the image in Figure 10.11, we have the total estimate of weight as calculated above is 1488084.78 N and it is a six-laned road. So, estimate of average weight per lane is 248014.13 N. Assuming the stretch of road in Figure 10.11 has its length (l) thrice its breadth (b). Suppose the road needs to be widened to an eight-laned road. Then, the total weight the road has to handle at least $248014.13 \times 8 = 1984113.04$ N which can be rounded off to 2000000 N. The cross-sectional area for a similar stretch of road with same length and increased width can be calculated using $l \times \frac{b}{6} \times 8$ which is equivalent to $3b \times \frac{b}{6} \times 8$ where $b = 22.5$ m. Thus, the cross-sectional area is $4b^2$ which is equivalent to $4 \times (22.5)^2$ is $2025\ \mathrm{m}^2$. The minimum tensile strength can be thus calculated as $\sigma = \frac{w}{A}$ which is $\sigma = \frac{2000000}{2025} \frac{N}{m^2}$ which is 987.5643 Nm^{-2}. To improve accuracy of estimation and for better results, images can be captured at different times of a day and road strength can be estimated for each of these images and on this data, statistical algorithms can be applied which can more precisely estimate the tensile strength of the road.

Another application is the analysis of traffic at traffic hotspots, like which kind of and how many vehicles of that type pass through that area at particular times of a day, which can be used further for training purposes and improvements to traffic management systems.

10.5 Conclusion

The vehicle detection, identification and counting software uses YOLO (You Only Look Once) trained on COCO (Common Objects in Context) dataset capable of detecting 80 everyday objects. YOLO processes all S×S grid cells simultaneously and does not iterate through each grid cell. One of the limitations of this YOLO-based vehicle detection software is that it cannot accurately detect vehicles when they are too close. The other limitation includes that it cannot detect special vehicles like ambulances, jeeps, vans etc. The vehicle detection, identification and counting software finds applications in traffic control systems like over-speeding vehicle identification, identification of vehicles in no entry and pedestrian-only zone, and civil engineering fields like highway tolling and estimation of road strength.

References

[1] Huansheng Song, Haoxiang Liang, Huaiyu Li, and Zhe Dai. Vision-based vehicle detection and counting system using deep learning in highway scenes. *European Transport Research Review*, 11, 12 2019.

[2] Saiyed Umer, Ranjan Mondal, Hari Mohan Pandey, and Ranjeet Kumar Rout. Deep features based convolutional neural network model for text and non-text region segmentation from document images. *Applied Soft Computing*, 113:107917, 2021.

[3] N Kumar, M Gupta, D Gupta, and S Tiwari. Novel deep transfer learning model for covid-19 patient detection using x-ray chest images. *Journal of ambient intelligence and humanized computing*, pages 1–10, 2021.

[4] Shasha Liu, Xiaoyu Li, Mingshan Gao, Yu Cai, Rui Nian, Peiliang Li, Tianhong Yan, and Amaury Lendasse. Embedded online fish detection and tracking system via yolov3 and parallel correlation filter. *OCEANS 2018 MTS/IEEE Charleston*, pages 1–6, 2018.

[5] Joseph Redmon and Ali Farhadi. Yolov3: An incremental improvement. *ArXiv*, abs/1804.02767, 2018.

[6] Karlijn Alderliesten. Yolov3 - real-time object detection. *https://medium.com/analytics-vidhya/yolov3-real-time-object-detection-54e69037b6d0*, 2020.

[7] G. Arun Sampaul Thomas and G. Manisha. Discovery of compound objects in traffic scenes images with a cnn centered context using open cv. *ICTACT JOURNAL ON IMAGE AND VIDEO PROCESSING*, 10:2113–2117, November 2019.

[8] cvlib. *https://www.cvlib.net*.

[9] Keerti kulkarni. Object detection: Yolo. *https://medium.com/analytics-vidh ya/object-detection-yolo-fc6647ddd11f*, 2020.

[10] Alvise Susmel. Real-time object detection with phoenix and python. 2020.

[11] Alexey Bochkovskiy, Chien-Yao Wang, and Hong-Yuan Mark Liao. Yolov4: Optimal speed and accuracy of object detection. *arXiv preprint arXiv:2004.10934*, 2020.

[12] Ayoosh Kathuria. What's new in yolo v3. *https://towardsdatascience.com/yolo-v3-object-detection-53fb7d3bfe6b*, 2018.

[13] Detector loss function (yolo loss). *https://www.oreilly.com/library/view/hands-on-convolutional-neural/9781789130331/8881054c-f6e6-485c-9c9e-357285bce60a.xhtml.*

[14] Parneet Kaur, Yogesh Kumar, Shakeel Ahmed, Abdulaziz Alhumam, Ruchi Singla, and Muhammad Fazal Ijaz. Automatic license plate recognition system for vehicles using a cnn. *Computers, Materials and Continua*, 71:35–50, 11 2021.

11

An Extensive Study on Object Detection and Recognition using Deep Learning Techniques

Padmaja Grandhe[1], B. Swathi[2], K. S. R. Radhika[3], Srikanth Busa[4], and D. Omkar Aviansh[5]

[1]Department of Computer Science & Engineering, Potti Sriramulu Chalavadhi Mallikarjunarao College of Engineering & Technology, Kothapeta, A.P, India
[2]Department of Computer Science & Engineering, Lakireddy Balireddy College of Engineering, Mylavaram, AP, India
[3]Department of Computer Science & Engineering, Professor TKR College of Engineering and Technology, Hyderabad, Telangana, India
[4]Department of Computer Science & Engineering, Kallam Haranadhareddy Institute of Technology, Chowdavaram, Guntur, A.P
[5]Department of Computer Science and Engineering, Tagliatela College of Engineering, University of New Haven, West Haven, Connecticut, USA
E-mail: padmajagrandhe@gmail.com; buragaddaswathi@gmail.com; ksrradhika@tkrcet.com; srikanth.busa@gmail.com; domka1@unh.newhaven.edu

Abstract

During this pandemic situation, it is very important to maintain social distance between the people in crowded places. It is impossible for the management of shopping malls, cinema theatres, and other crowded places to observe social distance manually. In this paper, the proposed model studies different recognition algorithms for identifying persons through CCTV surveillance and checks the distance between objects and other persons through object detection algorithms. Object recognition is identifying the content surrounded by the boundary box to make the system get trained on the labels associated with objects. In contrast, object detection is a process of identifying the region in the image where objects exist and marking a rectangular boundary box along with that object. Object recognition is achieved using either machine learning models like a combination of Histogram Gradients and SVM, visual features are grouped as a bag and VJ algorithm, or deep learning models with the help of CNN. There are a few limitations to be taken care of by the recognition. Multi-classification of objects is a complicated process in fully connected neural networks, and the localization process has to identify the coordinates of the boundary box to get more accurate results. The limitations of the object detection process are associated with the computation of geometrical surface areas and volumes. This paper discusses object detection and recognition and their applications and different algorithmic approaches. It also surveyed various researches that are carried out on applications that are developed using these approaches.

Keywords: Object detection, object recognition, YOLO, RCNN, image classification, feature extraction, neural networks.

11.1 Introduction

In CCTV surveillance through the live video capturing and satellite images, the maintenance of crowded places is difficult through traditional image processing approaches. The model needs a faster and more advanced approach of deep learning [1] like "Computer Vision" to accurately measure the social distance between the objects and to recognize the persons frame by frame. Computer vision has its significance in object detection and recognition applications that wave back from the past many decades. It uses machine learning and trains the images with labeled data. The beauty of computer vision lies in the neural network architecture developed. The object detection process is associated with finding the location of the objects, and it classifies the labels associated with that object [2]. This process is illustrated in Figure 11.1

The traditional approach to detecting the objects is RCNN (region-based CNN). These regions are utilized to identify multiple objects in the image by displaying the class and coordinates of the object because general CNN architecture cannot identify the location of the object. It uses CNN to identify the visual features of the image [3], which gives training to the model [30–34] by using the machine learning algorithms on the annotated data. It is difficult to find the number of output variables in the initial stage only in the

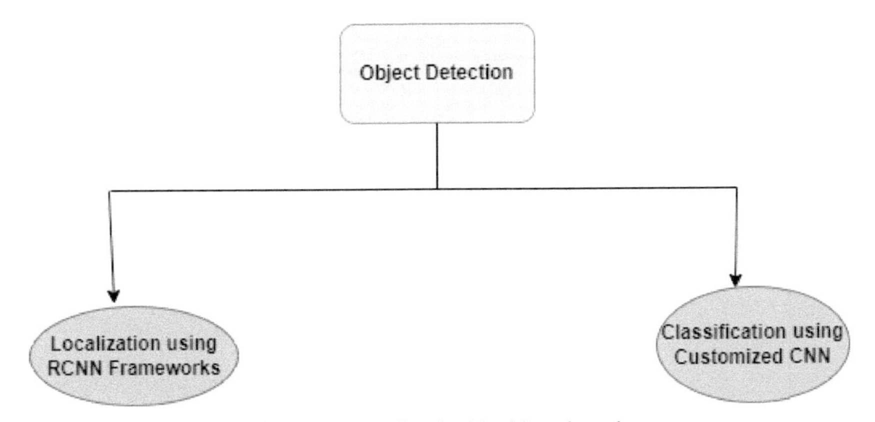

Figure 11.1 Steps associated with object detection process.

multi-object detection process, so the model cannot design a standard fully connected CNN for this purpose. This problem can be solved by identifying different regions necessary for the image and constructing a proposal region using the "Selective Search" approach integrating with RCNN. RCNN [35–37] forms four regions based on color, text, size, and locations [4]. These identified regions are customized as per the requirement of CNN for further classification. The entire process of RCNN is illustrated in Figure 11.2.

Initially, the input image is passed to the RCNN model, and then the selective search approach generates candidate regions by performing a sub-segmentation process. Then, to reduce the number of regions, it applies a greedy approach recursively to combine similar regions. Then combined regions are known as "Proposal Region." All the regions identified have different spatial coordinates, with which the CNN model cannot work because CNN needs a standard size as input. Hence, the model needs to apply geometric transformation functions like rescaling or resizing to get all the images of the same size [5]. Finally, the retrained model in which the last layers are designed using SVM for classification is applied as the last but one layer with its mathematical notation to minimize the distance between the points and find as many points as possible. This is shown in eqn 11.1 and 11.2.

$$P^* = arg_p(max(min(dist_H(\phi(X_n))))) \tag{11.1}$$

$$Dist_H(\phi(X_n)) = \frac{P^T(\phi(X_0)) + b}{\sqrt{P_1^2 + P_2^2 + P_3^2 + ... + P_n^2}}. \tag{11.2}$$

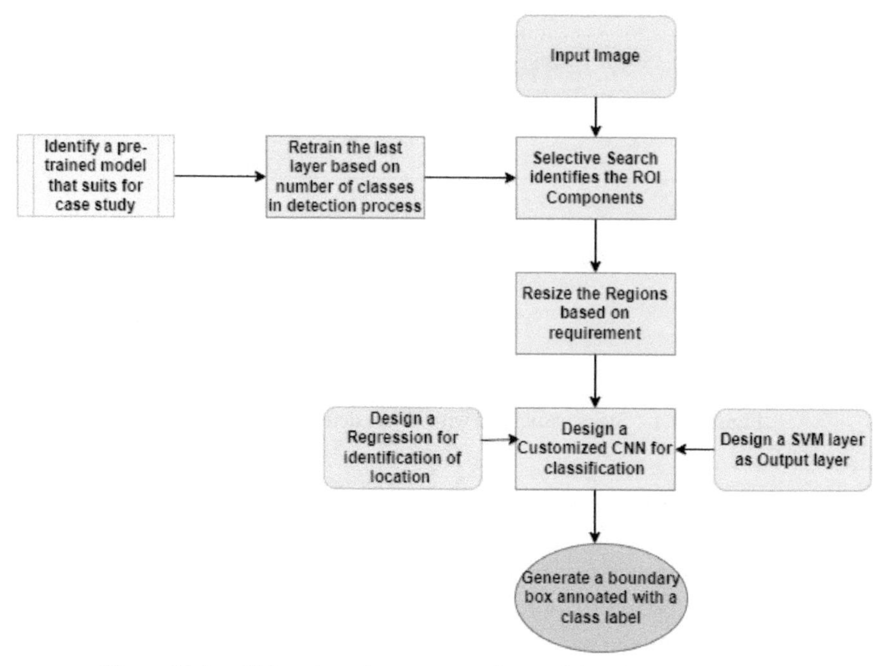

Figure 11.2 Object detection process using traditional RCNN approach.

Here, $P^T(\phi(X_0))+b$ represents a hyperplane equation to split the data points in the plane into positive and negative classes. The last layer in CNN is designed using the linear regression [6] to find the coordinates associated with the object using the eqn (11.3)

$$ind\_var_i = f(Feature_i, \beta) + Error\_Rate, \tag{11.3}$$

where $f(Feature\_i, \beta)$ represents a mapping function between the features of the image and unknown parameters. The below section discusses various applications of the detection mechanism.

1. Tracking an object plays a vital role in the security aspects by checking the live video streaming continuously
2. Counting of things or persons through images and videos to make the checking and packing as quickly as possible
3. Vehicle and person detection to identify things while unexpected things happen.
4. Identification of number plates at toll gates in the process of collecting toll passes, finding the rash driving to impose fines on particular

vehicles, to reduce the traffic at toll gates by automatic detection of FastScan tags, a major application of OCR

5. Self-driving cars are popular in abroad countries where cars can automatically identify the persons crossing the roads, signals at the traffic junction.

This object detection is obtaining popularity in terms of security areas, especially in restricted military areas to track non-living objects like drones and to keep on an eye on the movements of the living thing by continuously monitoring the prone areas within the borders of India [7]. The government of J&K can observe the real-time application of this object detection at their border positions. The military responds quickly in a fraction of seconds when it observes an unidentified object or person's movements in that particular region and takes necessary action. The object detection takes place by drawing a boundary box around the object it has identified, as shown in Figure 11.3.

Object detection can be achieved by performing traditional approaches or modern approaches like deep learning. The merits and demerits of both approaches are discussed in Table 11.1.

There are two types of object detection algorithms based on the concept of feature extraction, namely, two-stage algorithms [9] and one-stage algorithms. A two-stage process first identifies the number of objects, and then in the second step, it determines the size and marks a boundary. In a one-stage approach, it directly marks the boundary without involving in the

Figure 11.3 Object detection process using traditional RCNN approach.

Table 11.1 Pros and cons of traditional and modern approaches.

Approach	Description	Merits	Demerits
Traditional	This is an unsupervised approach [7] in which it does not require any training to make the system automate.	No need for additional resources to generate annotation images, which helps in labeling the data	The accuracy of the system is majorly affected by the noisy data and other uncontrolled factors.
Modern	This is a supervised approach [8], it trains the data using the computational power of GPUs	These types of systems can effectively handle both complex and dynamic scenarios. These produce the robust output within less time	More time is required to train the data, and more memory is required to produce the annotation data.

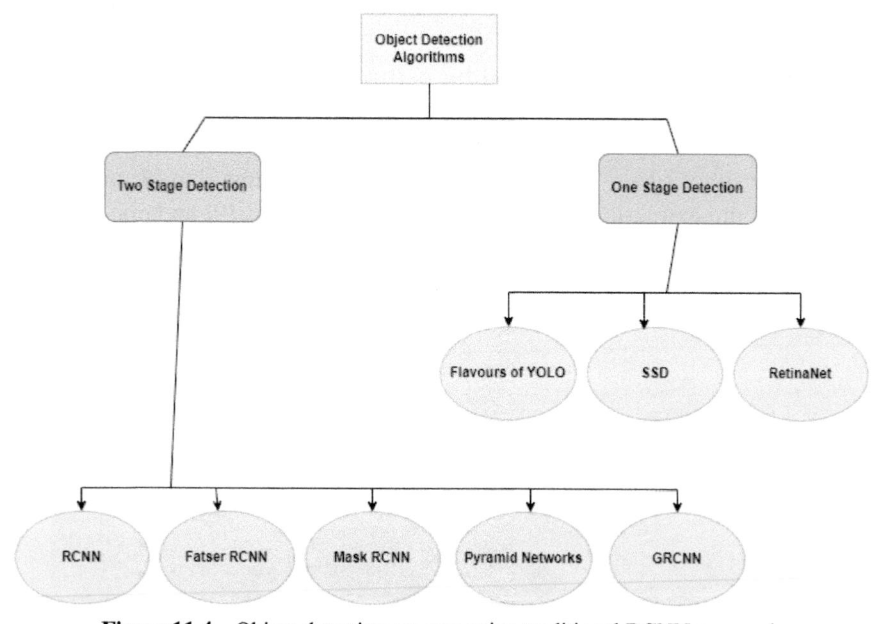

Figure 11.4 Object detection process using traditional RCNN approach.

process of region selection; this makes the entire process less complex. The classification of approaches is illustrated in Figure 11.4.

The two-stage detection mechanism first detects the object region using traditional approaches then object classification is performed based on the features extracted. It is good in terms of accuracy, but it is a very slow process because of n computations involved in region segmentation. In one stage of detection, it directly predicts the location without any sub-region computations, so this method is very fast and expensive. This method can also recognize linear and non-linear shapes, which two-stage detection algorithms cannot accomplish. The best algorithm for object detection in recent trends is "YOLO" (You Only Look Once), version 3. It is widely used in real-time object detection processes like person or objects detection in tracking systems. The main advantage of this system is that it is constructed based on a single neural network, which can efficiently divide the image into various grids. Each grid is of equal dimension, so the operation of resizing is reduced in this approach [10]. Every grid is associated with a class label, and it also computes the probability for each label, and then these candidate regions are combined to form a final prediction label based on the processing value. This architecture consists of 106 CNN layers and two fully connected layers [11]. Out of which, 53 layers are trained on ImageNet. The remaining 53 layers are customized based on the requirements of the case. The shape of the kernel is represented in the eqn (11.4)

$$Kernel_{\text{Shape}} = 1 * 1 * (B * (N + C)), \tag{11.4}$$

where B represents number of boundary boxes in the image. N represents number of attributes associated with the image. C represents number of class labels available. The entire process while working with YOLOv3 is illustrated in Figure 11.5.

The architecture layer for entire YOLOv3 for the pre-trained layers in the model is exhibited in Figure 11.6.

Object recognition is a two-step process that involves a combination of image classification, and localization [12]. During image classification, it applies any of the classification algorithms and predicts the class label among the available classes. The localization process involves drawing the boundary box along with the detected object. The output of the recognition is illustrated in Figure 11.7.

The major problem in object recognition occurs during the process of detection because most of the algorithms draw a rectangular boundary box around the object, which is very complicated to identify multiple objects which are residing side by side [13]. It cannot draw different polynomial shapes and exact boundaries for the identified shape. In the below section, Applications of object recognition are discussed.

1. Self-driving vehicles and robots are trending applications which attract the world of AI in recent years.

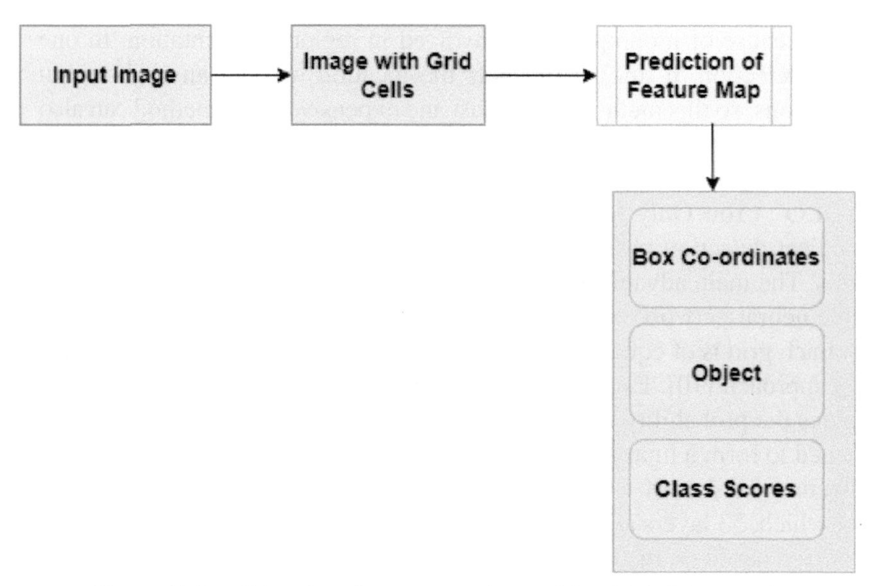

Figure 11.5 Object detection process using traditional RCNN approach.

2. In the medical field, the identification of severe disease based on the region of interest identification provides more accurate results about the stages of the disease rather than the doctors [14].
3. Face recognition, mask recognition are the major trending applications in the pandemic era for security purposes.
4. Image restoration, compression, reconstruction, and other image manipulation are the wider applications needed in cyber crime analysis.
5. Recognition of infected parts or portions from the CT Scan reports are the popular applications in the medical field [17].

For the model to recognize the elements, it must obtain sensor data like texture, color, shape, and motion. Then it has to get the descriptors associated with the image for doing easy analysis over the image [18]. If any image processing technique involves both detection and classification of images, then it is coined as "Recognition" [19]. The overall process is illustrated in Figure 11.8.

There are many approaches available for 2D and 3D object recognition. These classifications are presented in Figure 11.9.

In general, both the objects can be recognized using both traditional and deep learning approaches [20], but in the below figure, the chapter has discussed only the deep learning mechanisms.

11.2 Literature Review

In this research paper [21], the developer explains that objects' identification has witnessed various algorithm revisions to increase velocity and reliability performance. Through the constant efforts of several other investigators, DL algorithms with better detecting effectiveness are rising fast. The author elaborates on various real-time examples where object detection is applied by reducing the human effect due to the lack of ability in standard algorithms, which are said to be not quick and precise and are a tedious task to complete

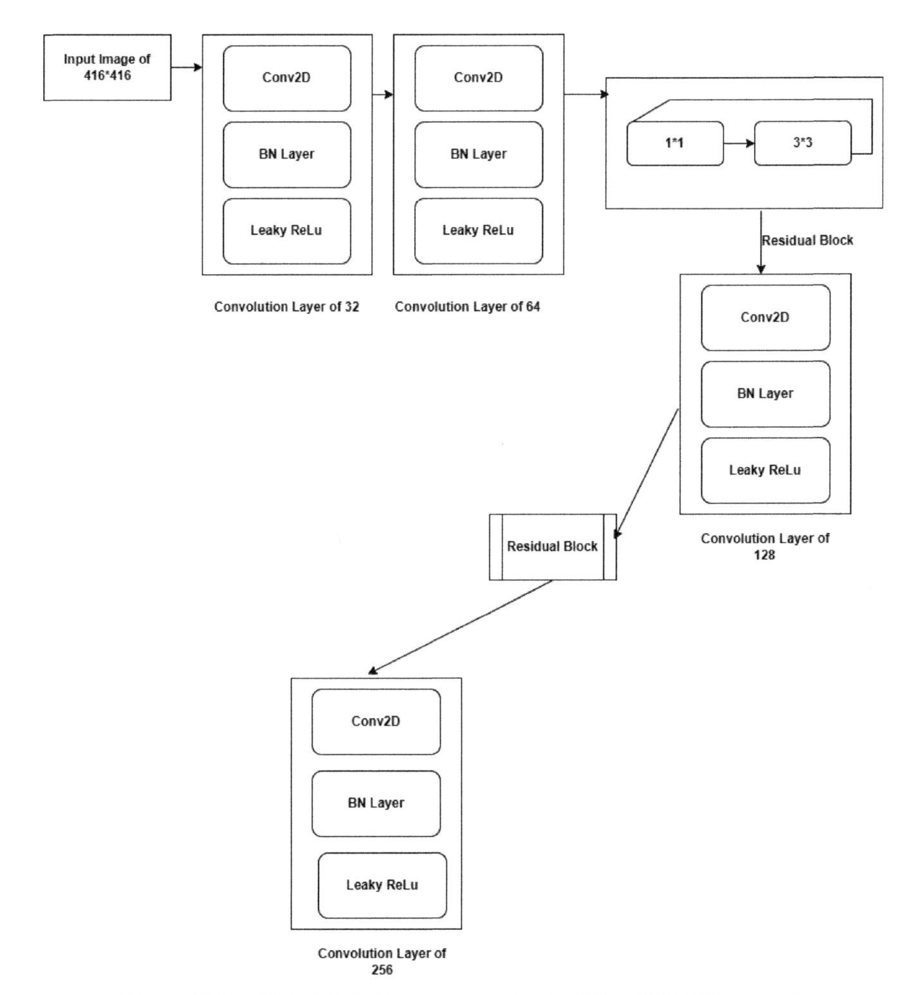

Figure 11.6 Object detection process using traditional RCNN approach.

Figure 11.7 Object detection process using traditional RCNN approach.

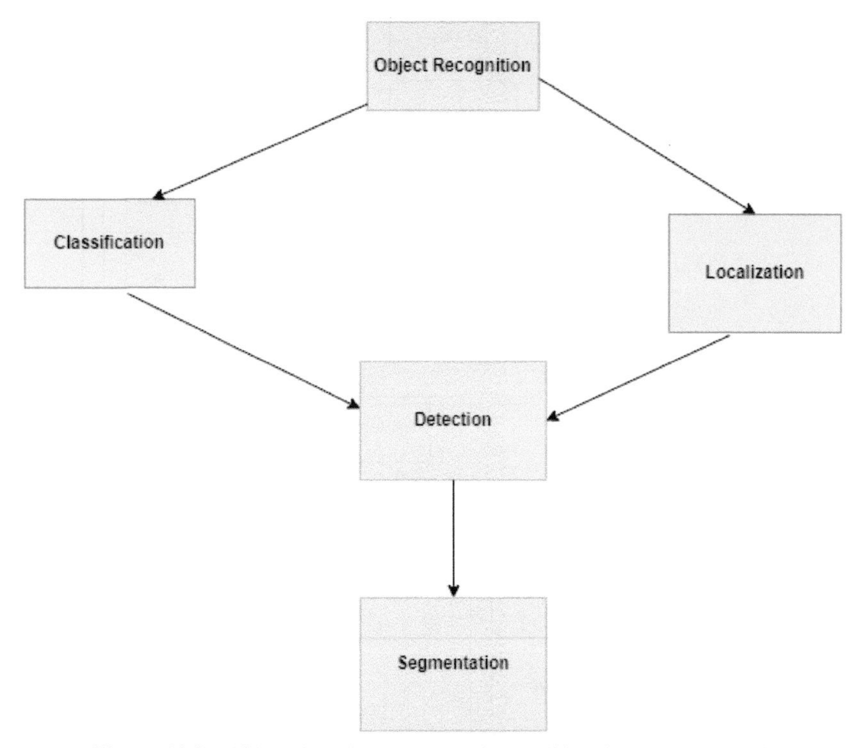

Figure 11.8 Object detection process using traditional RCNN approach.

all the issues at once. The main objective of this research work is to develop a machine that identifies the object in a picture both precisely and quickly.

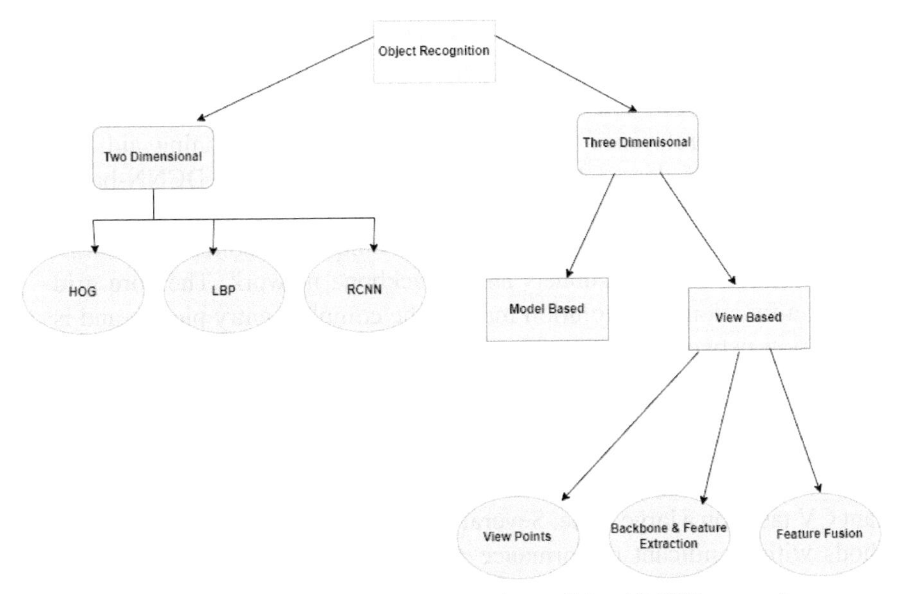

Figure 11.9 Object detection process using traditional RCNN approach.

Thus, they have developed a comparative study with two categories, namely one which performs RCNN in three substitute phases. At each phase, its algorithm is improved, and the fastness is increased. Various versions ranging from one to three of YOLO and SSD are performed in the second phase [22]. According to the author, recent studies grew more interested in object detection, and many frameworks were being developed in their respective study to make a more precise machine. Compared to a human, a machine lacks in identifying an object correctly and swiftly. Thus, this chapter proposes a study in identifying the objects using an advanced YOLOv1 NN. A YOLOv1 is an object identification within a single-stage model. Usually, this issue is a dependency problem to a spatially divergent enclosed box associated with the class probabilities. The entire process is done in a single evaluation. This module is further developed to lose its function of the YOLOv1 network and replaces the boundary style with proportion. By the author, the developed model showed reduced network error. The following are advancements made to the standard model: a pooling layer, namely, the spatial pyramid, and reduced parameters. 65.6% and 58.7% are the performance accuracies of the opted RCNN and YOLOv1, respectively [23]. Innovative developers of this research focused on implementing the recognition of interior objects pose a machine vision-based challenge to identify distinct ambient classes. In the recent few years, this job has received much interest. The high curiosity in this area can be understood by the significant relevance of this task for interior positioning aid and the advancement of deep CNN. This report tries with the DCNN-based architecture to construct a new downtown detector of objects. The developed structure is built on the DCNN "RetinaNet," a unique, consolidated network consisting of two mission subnets and a backbone network. The core grid calculates a functional convolution map on the complete entry picture and is a convolution network outside of itself. Evaluation is performed on various cores like DenseNet, VGGNet, and ResNet to enhance the durability and processing time and gained 84.61% mAP in identification accuracy [24]. By the author, the predominant that encourages human optical recognition is addressing the most appropriate items in a task that has been applied to variant CV tasks on a large scale. Several RGB-D-related object identification methods with significant performance were developed. However, recently, they have lacked a deeper understanding of these methodologies and complexities in this research area. This chapter is a rigorous study of this RGB-D framework under various aspects in finding its reactions towards each perspective over the chosen datasets. In addition, they have discussed important item identification algorithms and central standard databases from this discipline because light fields can also offer a depth map. Furthermore, a detailed indicator evaluation of different sample RGB-D-based, prominent object recognition approaches was conducted to explore the capacity of existent templates to recognize outstanding objects. In conclusion, RGB-D's complexities and solution paths for object identification were paved for future research [25]. This chapter focused on providing more security or advancements in identifying the objects for the passengers and their vehicles in different environments. The sensors used in this experiment detect the light and distance of the objects to the vehicles and directly retrieve their position and geometric shape of the object within the identification range. This research used the combination of complementary information gathered from the two sensors. Initially, 3D LiDAR data to manage precise item location. These attributes are later mapped onto the picture from the ROI of the method and are inputted to CNN. Through their size, the identification of these objects was identified by combining the properties of the last three layers of the CNN. The procedure finally shows: (1) 3D LiDAR offers approximately 86 true applicants per picture and a limited recall value of more than 95%, significantly reducing the separation period; (2) for each picture of the recommended procedure, the ordinary processing capacity is only 66.79 millisecond which fulfills the actual time of use, apart from rolling windows which make up lots of nominated object proposals. The framework showed 89.04% and 78.18% of identifying vehicles and individuals [26]. This research is concentrated on identifying illegal migrants to provide security measures to the citizens. Previously, RGB cameras were used, which lacked identifying objects at night and under irregular weather conditions. Thus, thermal cameras were introduced as a replacement for the RGBs. This research introduced an automatic identification of objects using CNN with heat cameras to replace the RGBs. A comparative study of standard frameworks with faster RCNN, SSD, Cascade RCNN, and YOLOv3 were performed. The study has experimented with the videos recorded at low lighting and various weather conditions at different scopes with different motions. The study states that YOLOv3 proved a faster outcome than the other detectors and was opted to continue the research further. The detection limits have been acceptable for the overall accuracy of 97.98% for all test cases. A small number of thermos photos have been employed for training. Furthermore, the findings of livestock identification in thermal photos were revealed, especially when it came to slip-around things and illegally bordering passages [15]. The author of this research paper explains the advancements in today's technology by introducing DL models to various real-time problems. These methodologies have their esteemed contributions in identifying objects in a picture to its least classification possible whose motive is to distinguish the objects to its lower levels as well. According to the developer here, such classification to its deepest levels is quite complex in performing at large within-a-class and within-all-the-classes variance. This research provides a complete view of several deep architectures and focusing on the developed frameworks' responsibilities. Initially, they have explained the functioning of the CNN and its levels following a briefing of divergent CNN frameworks from the standard procedures in LeNet to AlexNet, ZFNet, PNAS/ENAS, GoogleNet, VGGNet, Xception, ResNet, ResNeXt, SENet, and DenseNet. The objective of this research is to identify, small arms, wildlife, and human recognition [27]. For a variety of purposes, object tracking in overhead pictures is vital. The unpredictable alignment of items is the most problematic difficulty of this assignment, which is addressed by numerous profound learning strategies. Due to space knowledge scarcity, a dependency method for object localization in prior works has restricted reliability. The simulations struggle with the difference between the design and location of objects. This article proposes a framework that solves all the previous issues using a point-based detector. This methodology uses the spatial data exceptionally and functions on CNN to localize the selected point. The methodology disintegrates localization and identification of the respective pathways to overcome the miscalculations of the feature development. The instance-alignment step is engaged in the visual identification path to assure the synchronization between the characteristic map and orientated area. Overall, the point-based predictor may easily be integrated into the region-based detector and significantly enhance item classification. The framework shows a performance of 72.71 [28]. Domain responsive element identification with no supervision is intended to adjust detectors from a tagged source to an unlabeled objective domain. Most of the existing research uses a two-stage technique that initially produces regional recommendations and then discovers items of interest that are extensively used to reduce the inter-domain differences in both phases. Adverse training, however, can interfere with the convergence of examples that are well-matched by just harmonizing general proportions across divergent sectors. With the motive to solve this issue, the developers here have adopted a UaDAN framework that deals with the antagonistic constraint training to coordinate the high aligned and poorly aligned examples in their respective ways. They have created an ambiguity metric that evaluates the orientation of each specimen and adapts the degree of adverse training to records that are well-matched and badly positioned. We also utilize the ambiguity method to promote curricular learning, which first facilitates orientation at the object level and then gradually makes it harder to correlate instances. Table 11.1 exhibits the survey on different algorithms. Qi *et al.* [29] demonstrated various approaches involved in three-dimensional object recognition. The recognition of 3D objects using deep learning algorithms is gaining more attention in the medical field and virtual reality domain. The backbone of this work is the usage of pre-trained models integrated with matching algorithms. In the 3D object recognition model, points projection helps extract the features from the image. This model reduces the burden involved in the classification of 3D images. It uses a feature fusion mechanism by dividing the views into active and passive views to get the local and global descriptors accurately. The views are constructed using a pooling network as it is passed as an input to the CNN for the recognition process. Despite common 2D pre-trained models, 3D has ModelNet40, ShapeNetCore, and MIRO as pre-trained models for efficient construction of the model. Table 11.2 provides the merits and demerits associated with the existing models to identify which are the efficient models for 2D and 3D object detection and recognition process.

Table 11.2 Comparative study on deep learning algorithms.

SI.No.	Author	Algorithms used	Merits	Demerits and future work
1	P. Adarsh	RCNN, Fast RCNN, Faster RCNN, YOLOv1, v2, v3, and SSD.	Identification of an image at various dimensions was shown.	In the future, they aimed to focus on smaller objects.
2	Tanvir	RCNN, YOLOv1	Loss function from the standard mechanism is modified, spatial pooling layer is added	Lower accuracy of the developed model. In the future, a hybrid detector can be used for smaller objects identification.
3	Afif	RetinaNet	The developed model had shown tremendous results in learning the indoor environment.	The combinatorial approach may retrieve greater results in future enhancements with larger data.
4	Zhou	RGB-D	Several solution paths for respective complexities were given when performing with RGB-D were explained.	Stated that more improvements were to be needed in RGB-D even after this survey.
5	X. Zhao	Complimentary	The framework was developed in identifying the object shapes and individuals to provide safety for users in vehicles.	The further study focuses on identifying the objects at the non-overlapping areas of the sensors.
6	M. Krišto	Faster RCNN, SSD, RCNN, Cascade RCNN, YOLOv3	Overcame the limitations in using the RGB-D cameras and performed a rigorous study to identify the objects at various dimensions at irregular conditions.	In the future, the study could be expanded in identifying individuals in hot climatic conditions and wild animals.

Table 11.2 (Continued.)

SI.No.	Author	Algorithms used	Merits	Demerits and future work
7	Dhillon	LeNet,AlexNet, ZFNet, PNAS/ENAS, GoogleNet, VGGNet, Xception, ResNet, ResNeXt, SENet and DenseNet	A summary of several CNN techniques starting from its initial level to the deeper levels of processing is explained.	In the future, this research could be extended to recognition in videos. Optimizing the current issues could also be carried.
8	Z. Chang	Point-based detector	Recognition of comparatively smaller objects was achieved on a diversified image.	In the future, identification through a videotape can also be developed.
9	D. Guan	UaDAN	The framework accurately classified the objects irrespective of the weather conditions and uncertain data.	Further studies could explore identifying the domain flexible jobs in picture recognition and semantic distribution.

11.3 Conclusion

From the above surveys carried out by various researchers, this chapter has plotted the accuracies obtained in Table 11.3. Table 11.2 represents the merits and limitations that are caused by the previously proposed systems. Figure 11.10 clearly shows that the combinations of detection algorithms have achieved more accuracy than the remaining mechanisms.

The identification and recognition of objects with the usage of deep learning techniques have made a good improvement in providing efficient surveillance systems in the real world. Among the existing 2D and 3D object recognition mechanisms, 3D approaches provide a better way of retraining the pre-trained model as per the requirements of the case study. In future work, researchers can work on algorithms that can automatically identify the polynomial shape of the objects, especially in the case of crowded places because entities exist in multiple forms. So, the model needs more attention to irregular shapes than regular shapes. Researchers can also work on finding

Table 11.3 Workflow analysis on accuracies.

Si.No.	Author	Algorithms used	Accuracy obtained
1.	P. Adarsh	Detection algorithms	95%
2.	Tanvir	RCNN and YOLOv1	65.6%
3.	Afif	RetinaNet	84.61%
4.	Zhou	RGB-D	78.3%
5.	X. Zhao	Complementary 3D LIDAR	89.04%
6.	M. Krišto	Cascade RCNN and YOLOv3	97.9%
7.	Dhillon	Neural networks	79.8%
8.	Z. Chang	Point-based detector	72.71%
9.	D. Guan	UaDAN	89.07%

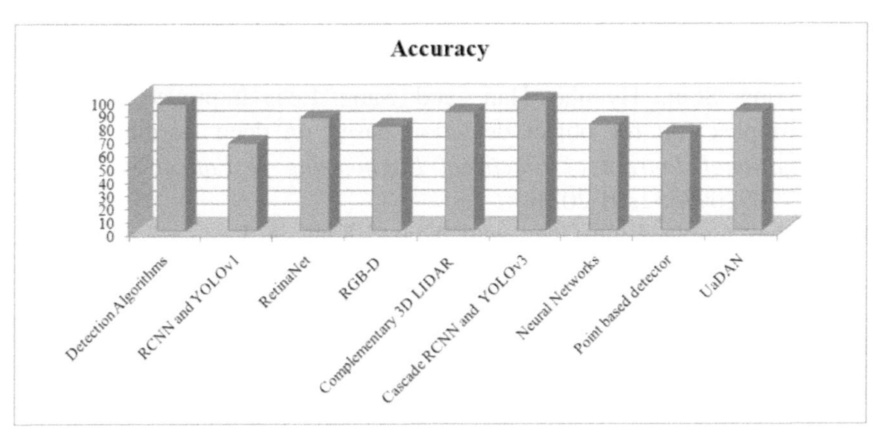

Figure 11.10 Object detection process using traditional RCNN approach.

good activation functions by combining the genetic approach with existing approaches to optimize the problem-solving process.

References

[1] Ning Wang, Yuanyuan Wang, and Joo Er. Review on deep learning techniques for marine object recognition: Architectures and algorithms. *Control Engineering Practice*, 118:104458, 05 2020.

[2] Riadh Ayachi, Yahia Said, and Mohamed Atri. A convolutional neural network to perform object detection and identification in visual large-scale data. *Big Data*, 9, 10 2020.

[3] Asra Aslam and Edward Curry. A survey on object detection for the internet of multimedia things (iomt) using deep learning and event-based

middleware: Approaches, challenges, and future directions. *Image and Vision Computing*, 106:104095, 2021.

[4] Sankar Pal, Anima Pramanik, Jhareswar Maiti, and Pabitra Mitra. Deep learning in multi-object detection and tracking: state of the art. *Applied Intelligence*, 51, 09 2021.

[5] Hirofumi Watanabe, Yoshiko Ariji, Motoki Fukuda, Chiaki Kuwada, Yoshitaka Kise, Michihito Nozawa, Yoshihiko Sugita, and Eiichiro Ariji. Deep learning object detection of maxillary cyst-like lesions on panoramic radiographs: preliminary study. *Oral radiology*, 37, 09 2020.

[6] Bo Xiao and Shih-Chung Kang. Development of an image dataset of construction machines for deep learning object detection. *Journal of Computing in Civil Engineering*, 35:05020005, 03 2021.

[7] Seokyong Shin, Hyunho Han, and Sang Lee. Improved yolov3 with duplex fpn for object detection based on deep learning. *The International Journal of Electrical Engineering & Education*, page 002072092098352, 01 2021.

[8] Abhishek Gupta, Ling Guan, and Ahmed Khwaja. Deep learning for object detection and scene perception in self-driving cars: Survey, challenges, and open issues. *Array*, 10:100057, 02 2021.

[9] Daniel Saavedra, Sandipan Banerjee, and Domingo Mery. Detection of threat objects in baggage inspection with x-ray images using deep learning. *Neural Computing and Applications*, 33:1–17, 07 2021.

[10] Du Jiang, Gongfa Li, Chong Tan, Li Huang, Ying Sun, and Jianyi Kong. Semantic segmentation for multiscale target based on object recognition using the improved faster-rcnn model. *Future Generation Computer Systems*, 123, 05 2021.

[11] Youshan Zhang and Brian Davison. Domain adaptation for object recognition using subspace sampling demons. *Multimedia Tools and Applications*, 80, 06 2021.

[12] Munish Kumar, Monika Bansal, and Manish Kumar. 2d object recognition techniques: State-of-the-art work. *Archives of Computational Methods in Engineering*, 28, 02 2020.

[13] Munish Kumar, Monika Bansal, and Krishan Saluja. An efficient technique for object recognition using shi-tomasi corner detection algorithm. *Soft Computing*, 25, 03 2021.

[14] Kriti Ohri and Mukesh Kumar. Review on self-supervised image recognition using deep neural networks. *Knowledge-Based Systems*, 224:107090, 04 2021.

[15] Du Jiang, Gongfa Li, Ying Sun, Jiabing Hu, Juntong Yun, and Ying Liu. Manipulator grabbing position detection with information fusion of color image and depth image using deep learning. *Journal of Ambient Intelligence and Humanized Computing*, 12:1–14, 12 2021.

[16] Ruhul Amin Khalil, Edward Jones, Mohammad Babar, Tariqullah Jan, Mohammad Zafar, and Thamer Alhussain. Speech emotion recognition using deep learning techniques: A review. *IEEE Access*, PP:1–1, 08 2019.

[17] Adedamola Adedoja, Owolawi Pius, and Temitope Mapayi. Deep learning based on nasnet for plant disease recognition using leave images. 08 2019.

[18] Shervin Minaee, Yuri Y Boykov, Fatih Porikli, Antonio J Plaza, Nasser Kehtarnavaz, and Demetri Terzopoulos. Image segmentation using deep learning: A survey. *IEEE transactions on pattern analysis and machine intelligence*, PP, February 2021.

[19] Zhi Yang, Wei Yu, Pengwei Liang, Hanqi Guo, Likun Xia, Feng Zhang, Yong Ma, and Jiayi Ma. Deep transfer learning for military object recognition under small training set condition. *Neural Computing and Applications*, 31, 10 2019.

[20] Pranav Adarsh, Pratibha Rathi, and Manoj Kumar. Yolo v3-tiny: Object detection and recognition using one stage improved model. pages 687–694, 03 2020.

[21] Tanvir Ahmad, Yinglong ma, Muhammad Yahya, Belal Ahmad, Shah Nazir, Amin Haq, and Rahman Ali. Object detection through modified yolo neural network. *Scientific Programming*, 06 2020.

[22] Mouna Afif, Riadh Ayachi, Yahia Said, Edwige Pissaloux, and Mohamed Atri. An evaluation of retinanet on indoor object detection for blind and visually impaired persons assistance navigation. *Neural Processing Letters*, 51, 06 2020.

[23] Tao Zhou, Deng-Ping Fan, Ming-Ming Cheng, Jianbing Shen, and Ling Shao. Rgb-d salient object detection: A survey. *Computational Visual Media*, 7, 01 2021.

[24] Xiangmo Zhao, Pengpeng Sun, Zhigang Xu, Haigen Min, and Hongkai Yu. Fusion of 3d lidar and camera data for object detection in autonomous vehicle applications. *IEEE Sensors Journal*, PP:1–1, 01 2020.

[25] Mate Krišto, Marina Ivašić-Kos, and Miran Pobar. Thermal object detection in difficult weather conditions using yolo. *IEEE Access*, PP:1–1, 07 2020.

[26] Anamika Dhillon and Gyanendra Verma. Convolutional neural network: a review of models, methodologies and applications to object detection. *Progress in Artificial Intelligence*, 9, 12 2019.

[27] Kun Fu, Chang Zhonghan, Yue Zhang, and Xian Sun. Point-based estimator for arbitrary-oriented object detection in aerial images. *IEEE Transactions on Geoscience and Remote Sensing*, PP:1–18, 09 2020.

[28] Dayan Guan, Jiaxing Huang, Aoran Xiao, Shijian Lu, and Yanpeng Cao. Uncertainty-aware unsupervised domain adaptation in object detection. *IEEE Transactions on Multimedia*, PP:1–1, 05 2021.

[29] Shaohua Qi, Xin Ning, Guowei Yang, Zhang Liping, Peng Long, Weiwei Cai, and Li Weijun. Review of multi-view 3d object recognition methods based on deep learning. *Displays*, 69:102053, 07 2021.

[30] Sri Padmanabhuni. An extensive study on classification based plant disease detection system. *JOURNAL OF MECHANICS OF CONTINUA AND MATHEMATICAL SCIENCES*, 15, 05 2020.

[31] Padmaja Grandhe. A novel method for content based 3d medical image retrieval system using dual tree m-band wavelets transform and multiclass support vector machine. *Journal of Advanced Research in Dynamical and Control Systems*, 12:279–286, 03 2020.

[32] Padmaja Grandhe, Sreenivasa Edara, and Vasumathi Devara. Adaptive roi search for 3d visualization of mri medical images. pages 3785–3788, 08 2017.

[33] Padmaja Grandhe, Sreenivas Edara, and Vasumathi Devara. Adaptive analysis & reconstruction of 3d dicom images using enhancement based sbir algorithm over mri. *Biomedical Research (India)*, 29:644–653, 01 2018.

[34] Padmaja Grandhe, Sreenivasa Edara, and Vasumathi Devara. *An Extensive Study of Visual Search Models on Medical Databases*, pages 219–228. 11 2018.

[35] Padmaja Grandhe, Edara Reddy, and D. Vasumathi. An adaptive cluster based image search and retrieve for interactive roi to mri image filtering, segmentation, and registration. 94:230–247, 12 2016.

[36] Saiyed Umer, Partha Mohanta, Ranjeet Rout, and Hari Pandey. Machine learning method for cosmetic product recognition: a visual searching approach. *Multimedia Tools and Applications*, 80, 11 2021.

[37] Saiyed Umer, Ranjan Mondal, Hari Pandey, and Ranjeet Rout. Deep features based convolutional neural network model for text and non-text region segmentation from document images. *Applied Soft Computing*, 113:107917, 09 2021.

12

A Comprehensive Review on State-of-the-Art Techniques of Image Inpainting

Sariva Sharma and Rajneesh Rani

Dr B R Ambedkar National Institute of Technology, India
E-mail: sariva03@gmail.com

Abstract

Image inpainting, or the reconstruction of antique and corrupted images, has been around for a long time. It is a functional computer vision research problem with the goal of improving an image quality by removing unwanted details, adding missing elements, and presenting the image in a way that appeals to the human visual system. Image inpainting techniques are frequently employed in image reconstruction, such as in museums that may not be able to afford to engage an expert to repair corrupted paintings, image compression, object removal, and so on. This study provides a detailed survey and comparative assessment of several inpainting techniques ranging from basic inpainting techniques to deep learning inpainting techniques. The utility of these techniques is presented with notable comparisons and evaluated by reviewing the various factors as well as available datasets that researchers might utilise to evaluate their suggested method.

Keywords: Image inpainting, object removal, PDE-based inpainting, exemplar-based inpainting, deep learning, digital images.

12.1 Introduction

Nowadays, information in the form of photos and videos is required in a variety of fields, and processing the content of images and videos has become a difficult task due to the large amount of data contained therein. Image processing techniques have been applied to a wide range of applications, including image inpainting, video processing, medical image processing, image compression, printing industry, and many others. Image inpainting is a popular image processing technique. It refers to the process of restoring damaged image portions by carefully using information from undamaged regions in accordance with certain principles, resulting in a more realistic image that a naive human would not notice the restoration traces [1]. The efforts are mostly focused on obtaining an inpainted image that closely mimics the original undamaged image. It is an ill-assorted problem because it has no exact and universal solution. Many solutions have been proposed based on various parameters to obtain optimal solution. Inpainting is now used in a variety of applications, including medical image corrections, the restoration of old and damaged historical photos at museums, the creation of special effects for movies and television shows, image coding and transmission, automatic text removal, photo editing, the detection of forged images, the removal of unwanted objects such as video logos, and so on [2–5].

Researchers have begun implementing several image inpainting approaches using deep learning network models in recent years as deep learning technology has advanced rapidly. Its fundamental idea is to train the network using a large number of real photos, so that the network can learn the feature distribution of original images and then use this network to repair the damaged area of the image, so that image inpainting can be achieved [6–9]. The following is the paper's main contribution:

1. Various conventional and deep learning-based image inpainting algorithms are discussed in this review study. Accordingly, several techniques are evaluated based on performance metrics.
2. A brief explanation of several datasets is also provided in this study.
3. Lastly, the research gaps and challenges of image inpainting are addressed.

12.2 Related Work

A study has been carried out and presented in this section to explore several image inpainting techniques. This will help the research in understanding the various image inpainting techniques already in use and will provide new study directions.

12.2.1 Image inpainting using traditional techniques

PDE-based inpainting, texture synthesis-based inpainting, exemplar-based inpainting, and hybrid inpainting are the four primary categories of traditional image inpainting techniques. The goal of these techniques is to restore damaged images in such a way that alterations are not obvious to the naked eye, which means that the inpainted image must be realistic. A basic overview of traditional inpainting techniques is presented in the following section.

12.2.1.1 PDE-based inpainting techniques

The first inpainting technique was a PDE-based inpainting technique. Its concept is to diffuse undamaged information from the input image to the damaged region along the direction of the isophote, resulting in a linear propagation structure [10]. PDE-based inpainting techniques are used to recreate the planner structure, which can be thought of as a 1D pattern containing object and line contours. It works best for non-textured images where the major attention is on the geometrical structure of the image with only a little missing section. This method does not work well with textured images or images with large gaps since it introduces a blurry effect. PDE-based approaches require a non-trivial iterative procedure (anisotropic diffusion), which is difficult to implement, resulting in a long inpainting process. In general, this method is not appropriate for images with regular patterns since it uses only the boundary region of the damaged region to inpaint, which is insufficient information to recognize a regular pattern or texture in an image. Table 12.1 shows a comparison of various PDE-based approaches.

Bertalmio *et al.* [11] were the first to develop a PDE-based method that fills in missing regions with information near their boundaries when the user picks the region to be restored. Because of the concurrent fill-in of sections with varied structures and textures, restoration is quick. The primary goal of this method is to automate digital inpainting by mimicking basic approaches employed by expert restorators.

Biradar *et al.* [12] introduced a simple median filter-based digital inpainting technique (nonlinear filter). Median estimates the maximum likelihood of location. As a result, the method diffuses the median value of pixels from the boundary region into the inpainted area. The image's

Table 12.1 Comparison of PDE-based inpainting methods.

Year	Researcher	Performance parameters	Advantages	Disadvantages	Dataset
2000	Bertalmio *et al.* [11]	CPU time and multiresolution	Automatic inpainting, user only need to specify the region to be inpainted	Fails for restoring large texture area	Images from public dataset
2013	Biradar *et al.* [12]	PSNR (44.564)	Better result than Bertalmio, remove artificial noise	Fails for restoring multiple texture images.	Homogenous or heterogeneous texture images
2019	G.Sridevi *et al.* [13]	PSNR (37.5151)	Preserve edges and avoid blocky and staircase effect		USC-SIPI image

edge information is preserved by the median filter. Both homogeneous and heterogeneous textures work well with the proposed approach.

Sridevi *et al.* [13] suggested a PDE-based method that uses two new strategies to fill gaps: fractional order nonlinear diffusion driven by differential curvature (DC) and fractional order variational model to reduce noise and blur. The drawbacks with earlier PDE-based techniques include that the results are unstable and provide blurry results, and the angle of curves is not conserved with a simple variational model. By relying on the entire image rather than just the neighborhood pixel values, fractional order derivatives preserve edges features and remove staircase and speckle effects. The image's intensity changes are successfully characterised using difference curvature (DC).

12.2.1.2 Texture synthesis-based inpainting techniques

Texture synthesis-based inpainting approaches use information from the surrounding area to fix the damaged area. PDE-based approaches repair damaged regions by copying pixel information from the undamaged region, but this method produces enormous texture patterns from small input texture patterns while maintaining the statistical features of the input texture. There are a variety of texture-based inpainting techniques available. They are distinguished by how well they maintain image continuity between the original and damaged pixels. This technique isn't appropriate for all types of images; for example, it doesn't produce effective inpainting results with natural photos because it uses a regular pattern and fails to manage borders and boundaries. These techniques are only effective for tiny areas of damage. There are two sorts of texture inpainting techniques:

1. Based on statistical characteristics: The statistical-based texture inpainting approach can successfully make big irregular/non-liner textures, but it fails to handle regular/liner textures, such as floor tile, brick walls, and so on.
2. Pixel-based approaches: In pixel-based methods, the damaged region is repaired using pixels from the input texture pattern. As a result, it delivers better results than statistical approaches, but the computation time is longer because inpainting is done pixel by pixel. Furthermore, it is incapable of handling bigger regular texture patterns.

Table 12.2 shows a comparison of texture-based approaches.

Heeger *et al.* [17] proposed a texture-repair method based on pyramid-based texture synthesis. The method requires digitised texture and noise images as input. The goal of the method is to make the noise image look like the input texture image. This is accomplished through the use of an image pyramid and the Match-Histogram function. The Match-Histogram is used to force the noise intensity distribution to match the texture intensity distribution. To begin, the target texture is used to create an analysis pyramid. The synthesis pyramid is then built from noise iteratively, with the Match-Histogram algorithm used to each of the sub-bands of the analysis and synthesis pyramids. The updated synthesis pyramid is then collapsed back into noise, and the noise and target texture are matched using the match-Histogram algorithm. Image pyramid is a sub-band transform that produces a set of patches (sub-bands) of various sizes that correspond to distinct frequency bands (thus the name pyramid).

Table 12.2 Comparison of texture-based inpainting methods.

Year	Researcher	Performance parameters	Advantages	Disadvantages	Dataset
1999	Alexei A. Efros *et al.* [14]	Width of context window (5,11,15,23 pixels)	Works well for wide range of textures	Fails to handle highly textured images, high computational time	Variety of texture images
2001	Alexei A. Efros *et al.* [15]	Patch size	Works well with textured image with improved stability, less computational time	Sometimes fails to determine correctly what patch to use	Variety of texture images
2001	Michael *et al.* [16]	Window size (5×5 window size is preferred)	Works well for smooth natural textures	Does not work well for texture with regular structure	VisTex

To avoid blurring, Igehy *et al.* [18] presented an extended version of [17] with the purpose of compositing back the original image into the noisy image according to the mask using a multi-resolution approach. Rather than using a pyramid representation, noise is adjusted by combining the original image with the desired texture image, which is controlled by a mask.

Efros *et al.* [14] present a non-parametric texture-based inpainting approach. This approach creates a huge pattern from an input pattern pixel by pixel. This technique is useful for a wide range of images with different frequency information since pixel level manipulation is performed. The technique starts with the first pixel and synthesises all of the pixels in a square window around it to capture local properties. The amount of unpredictability in the texture pattern is determined by the size of the window.

Efros *et al.* [15] proposed an enhanced version of [14] where big texture patterns are created through patch level modification rather than pixel level manipulation. Patches of the input image are synthesised to create a large pattern.

For creating some natural texture, Michael *et al.* [16] devised a texture-based inpainting approach. The WL approach [19] is extended in this chapter, which works well for a wide range of textures but fails for a few specific types. This approach works excellent for those textures that require a faster response time.

12.2.1.3 Exemplar-based inpainting techniques

The most extensively utilized technique for inpainting that can repair huge damaged/missing areas is the exemplar-based technique. It's an iterative method for repairing damaged images by utilizing spatial features from the undamaged region until the desired image is attained. Inpainting can be done with patch-based methods utilizing this technique [20]. Patch-based techniques fix damaged areas patch by patch. This damaged patch is replaced with the closest similar exemplar patch from the undamaged region. The steps are as follows (Figure 12.1):

1. Initially, the region to be repaired, also known as the target region, is identified, and the selected region is organised in an appropriate data structure.
2. The priority of patches is then determined using a priority function that specifies the order of filling. The damaged patch with the most information is given the highest priority and is filled first.

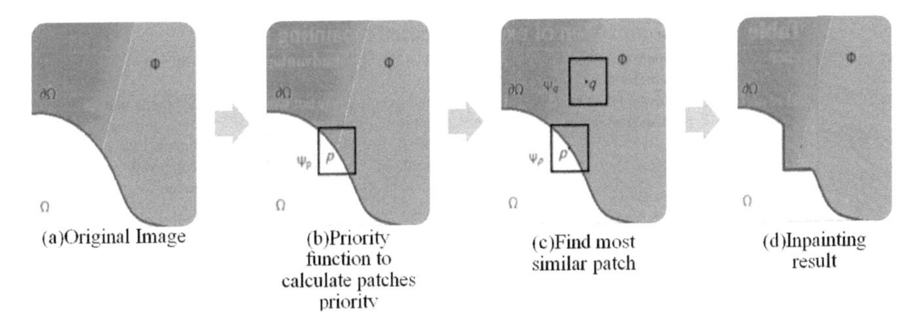

Figure 12.1 Procedure of exemplar-based techniques for image inpainting.

3. A similar patch to the one selected in the damaged region is searched for in the image's undamaged section.
4. Update the damaged region using similar patch in undamaged region.

Exemplar-based techniques perform well with rich texture images, realistic images that contain both texture and structure, and provide better results even for huge missing regions by minimizing blur. When compared to texture-based inpainting approaches, it reduces processing time. We can conclude from the approach of exemplar-based inpainting that it is a greedy algorithm, and hence it suffers from the fundamental issues of greedy techniques. The main downside of this strategy is that it fails to restore the damaged region if a significant amount of undamaged region is not available.

To achieve good inpainting results, both texture-based and patch-based approaches are required. When a high priority patch is used in patch-based approaches, overlapping with previously synthesized regions may occur while patch synthesizing to the damaged region. The already synthesised region is updated with the new patch due to overlaps. Texture-based approaches are utilized to avoid overlapping. Table 12.3 shows a comparison of exemplar-based techniques.

The idea of removing objects from images using exemplar-based methods was first introduced by Criminisi *et al.* in 2003 [28]. The core of this algorithm is an isophote-driven image sampling process; isophote refers to the direction and intensity of the patch center point. A patch is a small block (mostly rectangular) in the image. Generally, in digital image inpainting, Ψ_p refers to the target patch in which pixel values are missing and Ψ_q refers to the best match patch found in the source image. The important step is to assign priorities to patches in the target region, which determines the order in which they are filled. Patch priorities P(p) are determined by the confidence value

Table 12.3 Comparison of exemplar-based inpainting methods.

Year	Researcher	Performance parameters	Advantages	Disadvantages	Dataset
2004	Criminisi *et al.* [21]	Results are compared with original image	Edge sharpness is preserved; improve accuracy and execution speed	Does not work with curvature structures; If similar patches do not exist, the results are poor	Synthetic images and color images that include composite textures
2015	Deng [22]	Computation time comparison with previous methods	Work well with curved and cross-shaped structure	Not work well for geometry changed direction strongly i.e., corner of triangle	im2gps
2017	Ying *et al.* [23]	PSNR (46.49)	Robust; better inpainting quality	Image with distinct color and texture does not perform well	Image with rich texture and structure
2018	Yuheng *et al.* [24]	PSNR (42.1535) SSIM (0.9989)	Preserve edge structure and texture information	Long repair time because of use of global repair strategy	Random images
2018	Awati *et al.* [25]	Better results in term of SSIM (0.9824)	Preserve edges and corner information	Sometime not efficient with corners	Random images with high structural information
2020	Ahmed *et al.* [26]	PSNR (36.506)	Reduced the time it takes to find the best equivalent patch; high inpainting quality	Overhead for distance calculations	Natural and complex images of different file formats (JPG, PNG)
2020	Chang *et al.* [27]	DBP (Differen ces between patches), SSD (sum of squared differences)	Prevent the mismatch error		BSDS dataset

of surrounded pixels $C(p)$ and the data term $D(p)$.

$$P(p) = C(p)D(p). \tag{12.1}$$

The confidence value $C(p)$ represents the amount of information surrounding the patch's center pixel, i.e., texture information, whereas the data term $D(p)$ represents structure information. This is an iterative algorithm. In each iteration, once the best patch Ψ_q is found in the source region, it will be used immediately to fill in the missing part. This algorithm iterates until all patches are filled in the missing region.

Criminisi *et al.* in 2004 [21] proposed an extension of [28] that increases efficiency. Essentially, this work presents a more detailed description of the algorithm defined of [28] and a thorough comparison with the state of the art.

Deng *et al.* [22] proposed a novel priority function for choosing a target patch. This priority function first propagates geometry, which is represented by the data term $D(p)$, and then synthesizes image texture, which is represented by the confidence term $C(p)$. In this case, the priority function is divided into two phases, one based only on $D(p)$ and the other on $C(p)$, that can prevent image geometry and reconstruct image texture as well. The suggested approach increases image quality by reconstructing image geometry and texture.

Ying *et al.* in 2017 [23], addressed the issue of choosing the patch window size, which is not optimal for images with a lot of texture information and might lead to incorrect calculation of the maximum priority target block. The revised approach presented to handle this problem reconstructs the image utilizing image segmentation and the curvature property of isophotes. Curvature features investigate the details of texture, which are utilized to calculate priority. Second, rather than examining the complete source patch, this technique looks for a matching patch in the segmentation region.

Based on a novel Criminisi Algorithm, Yuheng and Hao *et al.* in 2018 [24] suggested an enhanced inpainting method. The approach focuses on two important features of any exemplar-based method, namely, it redefines the formula for priority calculation using a Manhattan distance and for matching of patch Euclidean distance is introduced. The original Criminisi Algorithm's priority calculation is based on the confidence term $C(p)$ of local pixels, which causes matching error that influences the repair order. As a result, the confidence term is replaced with the mean value of Manhattan distance.

Rao *et al.* [25] in 2018 proposed a new technique based on exemplar-based image inpainting that increases the quality of image structure, such as edges and corners. For obtaining the data term $D(p)$ in patch priority, this method employs fractional derivative and extra curvature finding methods. The remainder of the technique is the same as outlined by the original Criminisi Algorithm [28].

Ahmed *et al.* in 2020 [26] improved the searching mechanism of exemplar-based inpainting. Criminisi was the first one who proposed an inpainting method for the removal of objects or filling in missing areas using an exemplar-based technique. This technique is divided into two steps: (i) In the first step, the algorithm decides which patch has the highest priority to be filled first. (ii) The second step is the searching mechanism to discover the most similar patch to the selected highest priority patch. The main aim of this chapter [26] is the second step. To increase the speed of the searching mechanism, the sum of Euclidean and Position/Location distance is used to find the similarity between the highest priority patches and each patch of the input image. These steps are repeated until the last patch from the input image is checked.

12.2.1.4 Hybrid inpainting techniques

Hybrid inpainting techniques combine PDE-based and texture synthesis-based inpainting to restore the damaged region since images can contain both structure and texture, necessitating the usage of specialized inpainting techniques. The procedure for these techniques is as follows (Figure 12.2):

1. To begin, the input image is divided into two regions: texture and structure.
2. Both areas are fixed independently with appropriate inpainting techniques.
3. At last, both repaired regions are combined to get final results.

Table 12.4 shows a comparison of hybrid inpainting approaches.

Bertalmio *et al.* in 2003 [29] presented a hybrid inpainting approach. The original image is divided into two independent images, texture and structural image, using this technique. The Bertalmio approach [11] is used to repair

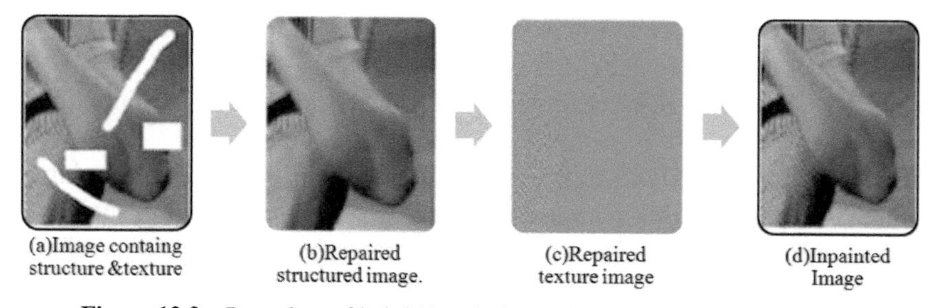

Figure 12.2 Procedure of hybrid inpainting techniques for image inpainting.

Table 12.4 Comparison of hybrid inpainting methods.

Year	Researcher	Performance parameters	Advantages	Disadvantages	Dataset
2003	Bertalmio *et al.*	Computational time	Outperform as compared to separate use of texture and inpainting technique.	Does not work for video and 3D images.	Random images contains structure and texture.
2015	Biradar *et al.*	PSNR(22.236)	Fast and works better for large objects	Works only for homogeneous region; Expensive approach	
2019	Kumar *et al.*	PSNR(25.86dB) SSIM(0.995dB)	Works well for irregular damaged region	—	TUM-IID

the texture image, and the Efros technique [14] is used to repair the structure picture. Finally, the results are combined to get a final result.

Biradar *et al.* [30] suggested a method for removing a huge item from a picture. Here, the Gaussian pyramid and recursive median approaches are merged. The Gaussian pyramid approach is in charge of minimizing the damaged region, while the recursive median technique fills it. Separately, the Gaussian pyramid can remove only small damaged objects, but the recursive median can repair only big portions while keeping the edge information, and this technique gets slow as the damaged region grows larger. It takes less time to remove a huge object after combining both procedures.

12.2.2 Image inpainting using deep learning

Use of deep learning technologies with image inpainting techniques is in demand because of its outstanding performance. The image inpainting procedure has become self-executing with the use of deep learning techniques that it does not require human involvement at all i.e., send an input/defected image to deep neural network and get a repaired image as an output or remove any objects that aren't necessary from the image quickly. In this chapter, we reviewed image inpainting approaches based on convolutional neural network (CNN) and generative adversarial network (gan) of deep learning models. The inpainting techniques based on these models such as Pathak *et al.* [31], Iizuka *et al.* [32], Wang *et al.* [33], Jo *et al.* [34], Nazeri *et al.* [35], Zheng *et al.* [36], and Li *et al.* [37] give state-of-the-art results in most of the cases. In these methods, the CNN is used as a feature extractor through the use of the convolution operation. The combination of CNN with GAN have achieved great results for image inpainting tasks [38] as CNN has an encoder that captures high-dimensional features, and a decoder that reconstructs images using the encoder's learnt features whereas GAN is responsible for handling the boundary details therefore provides sharped image.

Deep learning-based inpainting techniques require a large image dataset to train the network and subsequently to test the learnt network. As a result, the method's performance is determined by the dataset used. Different types of inpainting techniques use different sorts of datasets, and each technique responds differently to each data set. These datasets contains images having wide range of texture and structural characteristics, which can be used to test the technique's accuracy. We reviewed the popular deep learning-based image inpainting techniques in the next section before giving a comprehensive description of the datasets used by image inpainting techniques.

12.2.2.1 Convolutional neural network

The development of computer vision tasks utilizing deep learning has been established and fine-tuned over time with the usage of the convolutional neural network (CNN) algorithm. A deep learning algorithm, CNN works by taking image as a input, set weights and biases of different features in the images so that features can be differentiated according to its importance. With sufficient training, a CNN is capable to learn and understand the spatial and temporal features/characteristics of image through the use of suitable filters and it can generate/adjust filters by itself. The amount of pre-processing required in case of CNN method is much less compared to alternative method. Table 12.5 shows a comparison of CNN-based approaches. The techniques for image inpainting based on CNN architecture are summarized below.

Deep learning algorithms for blind inpainting were introduced by Nian Cai *et al.* in 2017 [45]. Repairing a corrupted image without knowing where the broken pixels are is known as blind inpainting. By cascading three convolutional layers to learn mapping between corrupted and ground truth subimage pairs, the proposed method accomplishes blind inpainting using a fully convolutional neural network. The blind inpainting CNN is trained using stochastic gradient descent with standard backpropagation.

Yan *et al.* in 2018 [39] presented a CNN-based technique by adding a Shift-Net layer to the U-net design. With the addition of this layer, images of any shape with sharp structures and fine-detailed textures can be filled without causing blur in the output images. Shift-Net makes use of both exemplar-based and CNN-based approaches. Shift-Net is trained using guided, adversarial, and reconstruction losses. The guiding loss is used to reduce the distance between the encoded feature in the known region and the decoded feature in the missing region.

For image inpainting, Zeng *et al.* in 2019 [46] introduced PEN-Net. Internally, the PEN-Net architecture employs a U-Net architecture. With U-Net structure image is repaired by encoding the context from low-level pixels to high-level semantic features and decoding the learned features back into images. The pyramid encoder's goal is to learn region empathy from high-level features learned and then transfer those features to low-level features using the pyramid structure.

To improve inpainting results, Ma *et al.* in 2019 [47] presented a two-stage coarse-to-fine inpainting approach. A region-wise convolution is conducted in the first stage to locally explore different sorts of areas using different convolution filters, which aid in the repair of missing regions from

Table 12.5 Comparison of CNN-based inpainting methods.

Year	Researcher	Mask/Image to mask resolution	Performance parameters	Resolution	Dataset
2018	Yan *et al.* [39]	Regular	PSNR (26.51), SSIM (0.90), L2 Loss (2.08)	256×256	ParisStreetView
2018	Liu *et al.* [40]	Irregular (1–10%) Irregular (10–20%) Irregular (20–30%) Irregular (30–40%) Irregular (40–50%) Irregular (50–60%)	PSNR (33.75), SSIM (0.94), L1 Loss (0.49) PSNR (27.71), SSIM (0.87), L1 Loss (1.18) PSNR (24.54), SSIM (0.77), L1 Loss (2.07) PSNR (22.01), SSIM (0.68), L1 Loss (3.19) PSNR (20.34), SSIM (0.58), L1 Loss (4.37) PSNR (18.21), SSIM (0.47), L1 Loss (6.45)	256×256	Places2
2019	Xie *et al.* [41]	Irregular (10–20%) Irregular (20–30%) Irregular (30–40%) Irregular (40–50%)	PSNR (28.51), SSIM (0.87), L1 Loss (1.12) PSNR (25.59), SSIM (0.78), L1 Loss (1.93) PSNR (23.31), SSIM (0.70), L1 Loss (2.55) PSNR (21.66), SSIM (0.60), L1 Loss (3.67)	256×256	Places 365
2020	Liu *et al.* [42]	Regular Irregular (10–20%) Irregular (20–30%) Irregular (30–40%) Irregular (40–50%)	PSNR (26.32), SSIM (0.91), FID (25.51) PSNR (31.13), SSIM (0.96), FID (6.91) PSNR (28.87), SSIM (0.92), FID (8.06) PSNR (25.34), SSIM (0.85), FID (19.36) PSNR (22.81), SSIM (0.71), FID (28.79)	256×256	CelebA Place2
2020	Saban *et al.* [43]	Missing pixels	PSNR (7.5), SSIM (0.42) PSNR (6.2), SSIM (0.33) PSNR (7.3), SSIM (0.36)	—	COREL(RGB) COREL (Gray) IRMA
2021	Wang *et al.* [44]	Irregular (10–20%) Irregular (20–30%) Irregular (30–40%) Irregular (40–50%) Irregular (10–20%) Irregular (20–30%) Irregular (30–40%) Irregular (40–50%) Irregular (10–20%) Irregular (20–30%) Irregular (30–40%) Irregular (40–50%)	PSNR (32.17), SSIM (0.91), L1 Loss (0.60) PSNR (27.96), SSIM (0.82), L1 Loss (1.33) PSNR (25.86), SSIM (0.76), L1 Loss (1.88) PSNR (24.28), SSIM (0.67), L1 Loss (2.59) PSNR (31.09), SSIM (0.89), L1 Loss (0.74) PSNR (27.52), SSIM (0.80), L1 Loss (1.53) PSNR (25.68), SSIM (0.73), L1 Loss (2.03) PSNR (23.74), SSIM (0.65), L1 Loss (2.91) PSNR (33.47), SSIM (0.92), L1 Loss (0.50) PSNR (28.66), SSIM (0.83), L1 Loss (1.13) PSNR (27.06), SSIM (0.78), L1 Loss (1.57) PSNR (24.62), SSIM (0.70), L1 Loss (2.35)	256×256	ParisStreetView Places2 CelebA-HQ256

existing regions. Following the first step, image inconsistencies develop due to a failure to examine the global association between existing and missing regions. As a result, a non-local operation is used in the second step to model correlation between distinct locations.

12.2.2.2 Generative adversarial network

Generative adversarial network (GAN) proposed in 2014 by Goodfellow *et al.* [38] and it falls under the category of generative models i.e., GAN models has capability to create and generate new data that could have been derived legitimately from the source dataset. The standardized GAN model is introduced by Radford *et al.* [48] in 2015 called as deep convolutional generative adversarial networks (DCGAN) and most of the GAN techniques are based on DCGAN. Deep learning architecture such as CNN are the basis of GAN models. The basic architecture of GAN consists two sub-models: generator and discriminator model. A generator model is used/trained in order to generate new data and a discriminator model examines whether the generated data is real (from the dataset) or fake (generated). These models are trained until the discriminator model is tricked roughly half of the time, indicating that the generator model is producing realistic data. Table 12.6 shows a comparison of GAN-based approaches. The techniques for image inpainting based on GAN architecture are summarized below.

Pathak *et al.* in 2016 [31] suggested an inpainting method based on context encoders from convolutional neural networks (CNNs). This is an algorithm for learning unsupervised visual features. The encoder–decoder architecture is used. Encoding the context of the target image patch and using the feature of image patch to get nearest contexts from a dataset yields patches that are semantically equivalent to the input patch on the encoder side. Tune the encoder for understanding tasks like categorization, object detection, and semantic segmentation to evaluate the learnt feature representation approach. Generate image pixels using the feature representation learned by the encoder at the decoder side. The proposed approach uses a channel-wise completely linked layer to connect encoder and decoder features.

Sagong *et al.* in 2019 [60] presented a method for inpainting that uses a parallel network (PEPSI) to reduce the number of operations by using a single-stage encoder–decoder network instead of a two-stage coarse-to-fine network. A single encoding network and a parallel decoding network are included in the PEPSI. Due to the parallel nature of the decoding network, it has two paths: coarse and inpainting. The coarse approach generates a rough inpainting result in which the encoder predicts features using a simple context

Table 12.6 Comparison of GAN-based inpainting methods.

Year	Researcher	Type	Mask/Image to mask resolution	Performance parameters	Resolution	Dataset
2016	Pathak *et al.* [31]	CE	Regular	PSNR (18.58), L1 Loss (9.37), L2 Loss (1.96)	227×227	ParisStreetView
2017	Yang *et al.* [49]	CE	Regular	PSNR (18.00), L1 Loss (10.01), L2 Loss (2.21)	128×128	ParisStreetView
2018	Yu *et al.* [50]	WGAN	Regular	PSNR (18.91), L1 Loss (8.6), L2 Loss (2.1)	256×256	Places2
2018	Demir *et al.* [51]	G-GAN & Patch-GAN	Regular	PSNR (19.03), SSIM (0.86), L1 Loss (5.54), L2 Loss (1.19)	256×256	ParisStreetView
2018	Wang *et al.* [33]	WGAN	Regular	PSNR (24.65), SSIM (0.86) PSNR (22.43), SSIM (0.89) PSNR (20.16), SSIM (0.86) PSNR (25.70), SSIM (0.95)	256×256	ParisStreetView ImageNet Places2 CelebA-HQ
2019	Yu *et al.* [52]	Patch-GAN	Regular Irregular	L1 Loss (8.6), L2 Loss (2.0) L1 Loss (9.1), L2 Loss (1.6)	512×512	Places2
2019	Liu *et al.* [53]	Patch-GAN	Regular Regular Irregular (10–20%) Irregular (20–30%) Irregular (30–40%) Irregular (40–50%) Irregular (10–20%) Irregular (20–30%) Irregular (30–40%) Irregular (40–50%)	PSNR (24.04), SSIM (0.82), L1 Loss (2.37), L2 Loss (0.52) PSNR (26.54), SSIM (0.93), L1 Loss (1.83), L2 Loss (0.27) PSNR (34.69), SSIM (0.98), L1 Loss (0.72), L2 Loss (0.04) PSNR (32.58), SSIM (0.98), L1 Loss (0.94), L2 Loss (0.07) PSNR (25.32), SSIM (0.92), L1 Loss (2.18), L2 Loss (0.37) PSNR (24.14), SSIM (0.88), L1 Loss (2.85), L2 Loss (0.44) PSNR (32.67), SSIM (0.97), L1 Loss (1.05), L2 Loss (0.08) PSNR (30.32), SSIM (0.95), L1 Loss (1.41), L2 Loss (0.13) PSNR (24.85), SSIM (0.87), L1 Loss (2.69), L2 Loss (0.45) PSNR (23.10), SSIM (0.76), L1 Loss (3.70), L2 Loss (0.68)	256×256	Place2 CelebA ParisStreetView
2019	Nazeri *et al.* [35]	Patch-GAN	Regular Irregular (0–10%) Irregular (10–20%) Irregular (20–30%) Irregular (30–40%) Irregular (40–50%) Irregular (50–60%)	PSNR (25.49), SSIM (0.89), L1 Loss (2.39), FID (1.90) PSNR (39.60), SSIM (0.98), L1 Loss (0.29), FID (0.20) PSNR (33.51), SSIM (0.96), L1 Loss (0.76), FID (0.53) PSNR (30.02), SSIM (0.93), L1 Loss (1.38), FID (1.08) PSNR (27.39), SSIM (0.89), L1 Loss (2.13), FID (1.80) PSNR (25.28), SSIM (0.85), L1 Loss (3.03), FID (2.82) PSNR (22.11), SSIM (0.77), L1 Loss (4.89), FID (5.30)	256×256	CelebA
2019	Sun *et al.* [54]	SRGAN	Regular Irregular	PSNR (19.78), SSIM (0.72), L1 Loss (5.52) PSNR (27.54), SSIM (0.92), L1 Loss (2.02)	256×256	VOC2017 COCO
2020	Chen *et al.* [55]	—	Regular	PSNR (22.68), SSIM (0.87)	—	Oxford Buildings
2020	Uddin *et al.* [56]	RaHinge GAN	Irregular (20–30%) Irregular (30–40%) Irregular (40–50%) Irregular (20–30%) Irregular (30–40%) Irregular (40–50%) Irregular (20–30%) Irregular (30–40%) Irregular (40–50%)	PSNR (28.89), SSIM (0.93), L1 Loss (1.20), L2 Loss (0.20) PSNR (28.67), SSIM (0.89), L1 Loss (1.44), L2 Loss (0.21) PSNR (26.11), SSIM (0.84), L1 Loss (2.23), L2 Loss (0.36) PSNR (27.15), SSIM (0.87), L1 Loss (2.01), L2 Loss (0.26) PSNR (25.03), SSIM (0.82), L1 Loss (2.78), L2 Loss (0.43) PSNR (23.88), SSIM (0.79), L1 Loss (3.21), L2 Loss (0.55) PSNR (28.05), SSIM (0.90), L1 Loss (1.55), L2 Loss (0.18) PSNR (27.53), SSIM (0.89), L1 Loss (1.73), L2 Loss (0.21) PSNR (25.45), SSIM (0.85), L1 Loss (2.25), L2 Loss (0.32)	512×680 256×256	Places365 ImageNet CelebA-HQ
2021	Zeng *et al.* [57]	—	Irregular	PSNR (36.66), L1 Loss (0.59), L2 Loss (0.10)	256×256	CelebA-HQ
2021	Zeng *et al.* [58]	AOT-GAN & Patch-GAN	Irregular (1–10%) Irregular (10–20%) Irregular (20–30%) Irregular (30–40%) Irregular (40–50%) Irregular (50–60%)	PSNR (34.79), SSIM (0.97), L1 Loss (0.55), FID (0.20) PSNR (29.49), SSIM (0.94), L1 Loss (1.19), FID (0.61) PSNR (26.03), SSIM (0.89), L1 Loss (2.11), FID (1.57) PSNR (23.58), SSIM (0.83), L1 Loss (3.20), FID (3.38) PSNR (21.65), SSIM (0.77), L1 Loss (4.51), FID (6.89) PSNR (19.01), SSIM (0.68), L1 Loss (7.07), FID (20.20)	512×512	Places2

Table 12.6 (Continued.)

Year	Researcher	Type	Mask/Image to mask resolution	Performance parameters	Resolution	Dataset
2021	Jam *et al.* [6]	WGAN	Irregular	PSNR (40.40), SSIM (0.94), L1 Loss (0.32), FID (3.09) PSNR (39.55), SSIM (0.91), L1 Loss (0.34), FID (17.64) PSNR (39.66), SSIM (0.93), L1 Loss (0.28), FID (4.47)	256×256	CelebA-HQ ParisStreetView Places2
2021	Zhao *et al.* [59]	CGAN	Irregular (0–20%) Irregular (20–40%) Irregular (40–60%) Irregular (60–80%)	FID (0.54) FID (1.69) FID (2.88) FID (4.13)	512×512	FFHQ

attention module (CAM). Simultaneously, the inpainting path generates a high-quality image with features anticipated by the encoder through the use of modified CAM.

Shin *et al.* in 2020 [61] proposed extended version of [60] by reducing the number of network parameters while maintaining the performance.

Demir *et al.* in 2018 [51] presented the PGGAN (Patch Global generative adversarial network) inpainting approach, which includes a discriminator that determines if a picture is real or fraudulent. PatchGAN assesses local details of a corrupted image partitioned into patches, whereas GlobalGAN evaluates the image as a whole. PGGAN combines PatchGAN and GlobalGAN to share network layers, then uses split paths to build two adversarial losses that capture local and global aspects of images.

Jiang *et al.* in 2020 [62] suggested a Wasserstein GAN-based inpainting algorithm that follows an autoencoder architecture. Jiang *et al.* offer a model that includes a generator, a global discriminator, and a local discriminator. They've also used skip-connection in the generator to improve the model's prediction power.

Image inpainting methods based on neural networks introduce structural distortions and blurred textures. Chen *et al.* in 2020 [63] proposed a generative network established using fusion model of squeeze-and-excitation networks deep residual (SE-DResNet) to take out the features of the image to get a clear image. The learned features have given importance by enhancing useful features and suppressing features that are not useful. To make the image look realistic, a contextual perception loss network has been added to constrain the similarity of local features.

Cai *et al.* in 2020 [64] suggested a GAN-based strategy to solve the constraint of obtaining only one optimal inpainting result while ignoring numerous other feasible outcomes. Style extractor is a Weiwei function that extracts style features from ground truth. Both the extracted style features and the ground truth are fed into the generator. The generator will learn how to map style features together. Finally, the model generates a large number of results that can be utilised to fill in enormous gaps.

Rather than dealing with a certain form of mask, Li *et al.* in 2020 [65] presented a GAN-based network (DeepGIN) to handle many sorts of masked images. DeepGIN is divided into two stages: coarse reconstruction and refining. The coarse reconstruction stage is used to make a rough estimate of missing pixels. To obtain a thorough description of missing pixels, the coarse reconstruction step's output is fed into the refinement stage. To handle varied types/sizes of masks, the spatial pyramid dilation algorithm is included in the network.

12.3 Performance Parameters

A subjective or objective comparison is carried out to compare the methodologies proposed by researchers. The subjective comparison relies on visual judgment, while the objective comparison relies on a variety of evaluation metrics. The following are the evaluation metrics used to assess the technique's efficacy:

1. MSE (mean squared error) [66] is an error metric used to determine the cumulative squared error between the original image and the inpainted image in order to evaluate the inpainted image's quality. The lower the value of MSE, the higher the image quality. It is calculated as follows

$$MSE = \frac{1}{RC} \sum_{c=1}^{R} \sum_{r=1}^{C} [S(r, c) - S'(r, c)]^2, \tag{12.2}$$

   where $S(r, c)$ is the original image, $S'(r, c)$ is the inpainted image, and r, c are the image's rows and columns.
2. The goal of the L1 Loss function is to lower the sum of the absolute difference between the original and inpainted image pixel values. The average error is reduced by using the L1 Loss function. It's calculated as follows:

$$L1\ Loss = \sum_{i=1}^{n} [\, Image_{\text{Original}} - Image_{\text{Inpainted}} \,]. \tag{12.3}$$

3. The goal of the L2 Loss function is to reduce the sum of squared differences (MSE) between original and inpainted image pixel values.

4. PSNR [67] is an error measure that determines the peak signal-to-noise ratio between the actual picture and the inpainted image in order to determine the quality of the inpainted image. When the PSNR is high, the inpainted image quality is high. It is calculated as follows:

$$PSNR = 10\log_{10}\frac{R^2}{MSE}, \tag{12.4}$$

where R is the maximum possible pixel value of the input image and MSE denotes the mean squared error.
5. SSIM (structural similarity index) compares the similarity of the original and inpainted images. The SSIM value ranges from 0 to 1. If the SSIM value is close to one, the images are similar or the quality of the inpainted image is good. To achieve an SSIM value close to one, the algorithm must reduce noise; thus, SSIM is a "superior quality metric" than PSNR.
6. Computational time: The amount of time necessary to complete the inpainting procedure. Less computational time is preferred.

12.4 Dataset

A variety of datasets are available for evaluating and comparing the performance of various image inpainting techniques. With the same inpainting technique, different image categories provide diverse results. Natural photos, artificial images, facial images, and a variety of other categories are included. The datasets used for inpainting are listed below Table 12.7:

1. Paris StreetView[1]: This collection was compiled using Google StreetView that includes street photos from cities all over the world. A total of 15,000 photos are included in the Paris street view.
2. Places[2]: There are 1.8 million photos in this dataset, divided into 365 scene categories. The validation set has 50 images per category, whereas the testing set contains 900 images per category. This dataset is used to train a model with a huge dataset using deep learning techniques.
3. ImageNet[3]: This is a big dataset arranged into a hierarchy, with each node containing 500 photos on average. There are over 14 million photos in the current edition of the collection.

[1]Private Dataset; Request Dataset: dpathak@cs.cmu.edu
[2]http://places2.csail.mit.edu/download.html
[3]http://image-net.org/

Table 12.7 Dataset used by image inpainting technique.

Dataset	Type of images	Resolution/Size of dataset	Year
Paris StreetView [68]	RGB	936×537/15K	2015
Places [69]	RGB	256×256/10 million	2017
ImageNet [70]	RGB	Various resolution/14 million	2015
Berkeley segmentation [71]	RGB and Grayscale	12K	2001
USC-SIPI image database [72]	RGB and Grayscale	256×256, 512×512, 1024×1024	1977
CelebFaces [73]	RGB	Various resolution/200K	2018

4. Berkeley segmentation (BSDS)[4]: In this dataset, 12,000 images segmented manually from 30 human subjects.
5. USC-SIPI image database[5]: This dataset is a collection of digitized images.
6. CelebFaces attributes dataset[6]: There are almost 200 thousand celebrity photos in this collection, each with 40 binary attribute annotations.

12.5 Findings

A overview of various image inpainting techniques is presented in this review work. Tables 12.1 to 12.6 reflect the results of this study's analysis depending on various parameters. The following are some of the findings:

1. Traditional inpainting techniques and deep learning-based inpainting techniques are the two broad categories of inpainting techniques described in this chapter. Different categories of traditional inpainting procedures have been shown to work effectively for different types of images. PDE-based approaches have produced excellent results for images with linear structures (Table 12.1), texture-based approaches can handle images with texture (Table 12.2), and the exemplar-based approach produces excellent results for images with structure and texture, as evidenced by performance parameters such as PSNR, SSIM,

[4]https://www2.eecs.berkeley.edu/Research/Projects/CS/vision/ bsds/

[5]http://sipi.usc.edu/database/

[6]http://mmlab.ie.cuhk.edu.hk/projects/CelebA.html

and others (Table 12.3). Traditional techniques, despite their benefits, have some flaws. These approaches, for example, do not work well with images that contain both boundaries and curves; inpainting time is slow, so they cannot handle large image datasets; introduces blur effect for large damaged regions; works well only for images that contain small damaged regions; and occasionally introduces artificial effects in the results. Deep learning-based techniques, on the other hand, have the ability to handle large datasets, as illustrated in Tables 12.5 and 12.6. These techniques can also repair damaged images with complex structures, textures, and edges because the model is trained on real-world images, the outcomes are more realistic than older methods. As a result, deep learning-based inpainting procedures are favored the majority of the time, but this does not rule out the employment of traditional methodologies. Traditional methods can also be used in some cases.

2. The comparison of inpainting techniques is based not only on performance criteria like PSNR and SSIM, but also on computational aspects like running time, training time, RAM size, coding language, image resolution, mask, dataset, and so on. Computational research is another key field for increasing image quality. It is concluded that if a comparison of multiple image inpainting techniques is required, computational parameter must also be considered. Tables 12.8 and 12.9 illustrate the results of a computational analysis of traditional and deep learning-based inpainting techniques. As seen in Tables 12.8 and 12.9, traditional inpainting approaches require a slower CPU with less RAM than deep learning-based inpainting techniques. Deep learning-based inpainting techniques rely on high-speed CPUs and GPUs with substantial memory requirements in order to handle massive datasets containing high-resolution images. As a result, while deep learning-based inpainting techniques perform well in terms of repairing damaged regions, they confront an issue in terms of computing complexity.
3. It has been noticed that inpainting techniques differ based on the sort of image used, the methodology used, and a variety of other factors. As a result, determining whether technique is better is challenging. As a result, we can conclude that image inpainting is an ill-sorted problem.
4. Because large datasets are readily available, some strategies perform well on one type of dataset while others work well on another. As a result, selecting optimal datasets for evaluating image inpainting techniques is impossible.

Table 12.8 Computational comparison of traditional inpainting techniques.

Technique	Processor	RAM/coding language	Image resolution	Dataset
PDE-based (2000) [11]	Pentium2	128Mb\C++	—	Public dataset
PDE-based (2019) [13]	Intel i3	4GB\MATLAB	128×128	USC SIPI image
Texture-based (2001) [16]	R10000 195 MHz	Nil\C	200×200	VisTex
Exemplar-based (2003) [28]	Pentium IV	1GB\nil	200×200	Color photos with rich textures, as well as synthetic images
Exemplar-based (2004) [21]	Pentium IV	1GB\Nil	200 × 200	Synthetic and color images that include composite textures
Exemplar-based (2015) [22]	i3-2370M	3.25GB\Nil	—	im2gps
Exemplar-based (2018) [24]	Intel Core i7	12GB\C++	512×512	Image with rich texture and structure
Exemplar-based (2020) [27]	Intel Core i7	4GB	—	BSDS

Table 12.9 Computational comparison of deep learning inpainting techniques.

Technique	Graphic processor	Features dimension	Learning rate	RAM/ language	Batch size	Image resolution	Mask	Dataset
CNN (2016) [31]	Titan X GPU	6×6×256 for both encoder and decoder	1×10^{-3}	—	—	256×256	Square	Paris Street View, ImageNet
CNN (2017) [45]	Intel Xeon E5-2640 v2	—	5×10^{-3}	128GB\ C++	1	512×512	—	Berkeley segmentation
CNN (2018) [39]	Titan X Pascal GPU	—	2×10^{-4}	—	1	256×256	Square	Paris StreetView
CNN (2018) [74]	NVIDIA GTX980	3×3×k where k represents input feature.	1×10^{-3}	64GB\ Python	1	256×256	—	Paris StreetView, ImageNet
CNN (2019) [47]	—	—	1×10^{-4}	—	48	256×256	Binary	Places
CNN (2019) [46]	TITAN V	—	—	—	—	256×256	Square	CelebA, Places
GAN (2018) [51]	NVIDIA Tesla P100	—	—	—	—	256×256, 512×512	Square	Paris Street View, Places
GAN (2020) [63]	RTX2080Ti-11 GPU	64×64×1024	4×10^{-3}	16GB\ Python	16	160×160	—	CelebA
GAN (2020) [62]	—	—	—	Python	—	256×256	Irregular	CelebA
GAN (2020) [64]	NVIDIA 1080	5×5×k	2×10^{3}	8GB\ Python	12	128×128	Square	CelebA
GAN (2020) [65]	NVIDIA GeForce	—	4×10^{-4}	—	4	256×256	Rectangular	CelebA

12.6 Research Gaps and Challenges

1. The major goal of image inpainting algorithms is to keep image quality high.
2. When an image comprises huge missing holes, texture, edges, and complicated patterns, the PDE-based technique fails and generates a blur effect. As a result, these algorithms are only utilized to repair tiny regions and require more computing effort; further research is required to inpaint the huge missing region efficiently.
3. Because exemplar-based methods require patch selection to fill in the missing region, and patch selection affects the inpainting process, an automatic procedure for patch selection will be necessary in the future.

The selection of a patch necessitates an entire search of patches in the source region; an efficient search technique could be investigated in the future. Although the basic exemplar-based technique only uses rectangular patches, non-rectangular patches may be investigated in the future.

4. Deep learning approaches require a significant amount of training time to prepare deep learning models. As a result, improvement is required to address this issue.

12.7 Conclusion

Image inpainting first appeared in the early centuries, and researchers are continuously working on improving its techniques due to its applications in a variety of fields, including the restoration of ancient damaged photos, the medical profession, image/video compression, and so on. We compared a variety of inpainting techniques in this review paper. The following is the conclusion:

1. More than 70 research publications connected to image inpainting techniques were studied in this work from 1999 to 2021, and a significant analysis of diverse techniques of image inpainting was mentioned. Researchers have shown many inpainting strategies based on traditional approaches and deep learning-based approaches.
2. This paper discusses traditional inpainting techniques such as PDE-based techniques, texture-based techniques, exemplar-based techniques, and hybrid techniques, as well as deep learning-based inpainting approaches. Deep learning approaches have been divided into two categories: CNN-based and GAN-based.
3. The analysis has been based on performance measures, and a comparison table of various image inpainting approaches has been presented. A table of comparisons of various strategies is also provided based on their computational parameters.
4. After completing a thorough investigation, it can be concluded that inpainting approaches based on deep learning are better suited for image restoration because models are built using big datasets and can handle complex scenes.
5. Finally, this work addressed a number of research gaps and issues that may be useful to the researcher in the near future.

References

[1] Qiang Guo, Shanshan Gao, Xiaofeng Zhang, Yilong Yin, and Caiming Zhang. Patch-based image inpainting via two-stage low rank approximation. *IEEE transactions on visualization and computer graphics*, 24(6):2023–2036, 2017.

[2] Xue-Cheng Tai, Stanley Osher, and Randi Holm. Image inpainting using a tv-stokes equation. In *Image Processing based on partial differential equations*, pages 3–22. Springer, 2007.

[3] Xianquan Zhang, Feng Ding, Zhenjun Tang, and Chunqiang Yu. Salt and pepper noise removal with image inpainting. *AEU-International Journal of Electronics and Communications*, 69(1):307–313, 2015.

[4] Pulkit Goyal, Sapan Diwakar, et al. Fast and enhanced algorithm for exemplar based image inpainting. In *2010 Fourth Pacific-Rim Symposium on Image and Video Technology*, pages 325–330. IEEE, 2010.

[5] Qin Jiang, Yang Chen, Guoyu Wang, and Tingting Ji. A novel deep neural network for noise removal from underwater image. *Signal Processing: Image Communication*, 87:115921, 2020.

[6] Jireh Jam, Connah Kendrick, Vincent Drouard, Kevin Walker, Gee-Sern Hsu, and Moi Hoon Yap. R-mnet: A perceptual adversarial network for image inpainting. In *Proceedings of the IEEE/CVF Winter Conference on Applications of Computer Vision*, pages 2714–2723, 2021.

[7] Mohamed Abbas Hedjazi and Yakup Genc. Efficient texture-aware multi-gans for image inpainting. *Knowledge-Based Systems*, page 106789, 2021.

[8] Chao Zhang and Tong Wang. Image inpainting using double discriminator generative adversarial networks. In *Journal of Physics: Conference Series*, volume 1732, page 012052. IOP Publishing, 2021.

[9] Mohamed Abbas Hedjazi and Yakup Genc. Image inpainting using scene constraints. *Signal Processing: Image Communication*, page 116148, 2021.

[10] Maria G Martini, Chaminda TER Hewage, and Barbara Villarini. Image quality assessment based on edge preservation. *Signal Processing: Image Communication*, 27(8):875–882, 2012.

[11] Marcelo Bertalmio, Guillermo Sapiro, Vincent Caselles, and Coloma Ballester. Image inpainting. In *Proceedings of the 27th annual conference on Computer graphics and interactive techniques*, pages 417–424, 2000.

[12] Rajkumar L Biradar and Vinayadatt V Kohir. A novel image inpainting technique based on median diffusion. *Sadhana*, 38(4):621–644, 2013.

[13] G Sridevi and S Srinivas Kumar. Image inpainting based on fractional-order nonlinear diffusion for image reconstruction. *Circuits, Systems, and Signal Processing*, 38(8):3802–3817, 2019.

[14] Alexei A Efros and Thomas K Leung. Texture synthesis by non-parametric sampling. In *Proceedings of the seventh IEEE international conference on computer vision*, volume 2, pages 1033–1038. IEEE, 1999.

[15] Alexei A Efros and William T Freeman. Image quilting for texture synthesis and transfer. In *Proceedings of the 28th annual conference on Computer graphics and interactive techniques*, pages 341–346, 2001.

[16] Michael Ashikhmin. Synthesizing natural textures. In *Proceedings of the 2001 symposium on Interactive 3D graphics*, pages 217–226, 2001.

[17] David J Heeger and James R Bergen. Pyramid-based texture analysis/synthesis. In *Proceedings of the 22nd annual conference on Computer graphics and interactive techniques*, pages 229–238, 1995.

[18] Homan Igehy and Lucas Pereira. Image replacement through texture synthesis. In *Proceedings of International Conference on Image Processing*, volume 3, pages 186–189. IEEE, 1997.

[19] Li-Yi Wei and Marc Levoy. Fast texture synthesis using tree-structured vector quantization. In *Proceedings of the 27th annual conference on Computer graphics and interactive techniques*, pages 479–488, 2000.

[20] Olivier Le Meur, Josselin Gautier, and Christine Guillemot. Examplar-based inpainting based on local geometry. In *2011 18th IEEE international conference on image processing*, pages 3401–3404. IEEE, 2011.

[21] Antonio Criminisi, Patrick Pérez, and Kentaro Toyama. Region filling and object removal by exemplar-based image inpainting. *IEEE Transactions on image processing*, 13(9):1200–1212, 2004.

[22] Liang-Jian Deng, Ting-Zhu Huang, and Xi-Le Zhao. Exemplar-based image inpainting using a modified priority definition. *PloS one*, 10(10):e0141199, 2015.

[23] Huang Ying, Li Kai, and Yang Ming. An improved image inpainting algorithm based on image segmentation. *Procedia Computer Science*, 107:796–801, 2017.

[24] Song Yuheng and Yan Hao. Image inpainting based on a novel criminisi algorithm. *arXiv preprint arXiv:1808.04121*, 2018.

[25] H Chinmayee Rao, Anupama Awati, Bhagyashree Pandurngi, and MR Patil. Image inpainting using exemplar based technique with improvised data term. In *2018 International Conference on Computational Techniques, Electronics and Mechanical Systems (CTEMS)*, pages 162–166. IEEE, 2018.

[26] Mariwan Wahid Ahmed and Alan Anwer Abdulla. Quality improvement for exemplar-based image inpainting using a modified searching mechanism. *UHD Journal of Science and Technology*, 4(1):1–8, 2020.

[27] Lei Zhang and Minhui Chang. Image inpainting for object removal based on adaptive two-round search strategy. *IEEE Access*, 8:94357–94372, 2020.

[28] Antonio Criminisi, Patrick Perez, and Kentaro Toyama. Object removal by exemplar-based inpainting. In *2003 IEEE Computer Society Conference on Computer Vision and Pattern Recognition, 2003. Proceedings.*, volume 2, pages II–II. IEEE, 2003.

[29] Marcelo Bertalmio, Luminita Vese, Guillermo Sapiro, and Stanley Osher. Simultaneous structure and texture image inpainting. *IEEE transactions on image processing*, 12(8):882–889, 2003.

[30] Rajkumar L Biradar. Hybrid approach for inpainting large object.

[31] Deepak Pathak, Philipp Krahenbuhl, Jeff Donahue, Trevor Darrell, and Alexei A Efros. Context encoders: Feature learning by inpainting. In *Proceedings of the IEEE conference on computer vision and pattern recognition*, pages 2536–2544, 2016.

[32] Satoshi Iizuka, Edgar Simo-Serra, and Hiroshi Ishikawa. Globally and locally consistent image completion. *ACM Transactions on Graphics (ToG)*, 36(4):1–14, 2017.

[33] Yi Wang, Xin Tao, Xiaojuan Qi, Xiaoyong Shen, and Jiaya Jia. Image inpainting via generative multi-column convolutional neural networks. *arXiv preprint arXiv:1810.08771*, 2018.

[34] Youngjoo Jo and Jongyoul Park. Sc-fegan: Face editing generative adversarial network with user's sketch and color. In *Proceedings of the IEEE/CVF International Conference on Computer Vision*, pages 1745–1753, 2019.

[35] Kamyar Nazeri, Eric Ng, Tony Joseph, Faisal Z Qureshi, and Mehran Ebrahimi. Edgeconnect: Generative image inpainting with adversarial edge learning. *arXiv preprint arXiv:1901.00212*, 2019.

[36] Chuanxia Zheng, Tat-Jen Cham, and Jianfei Cai. Pluralistic image completion. In *Proceedings of the IEEE/CVF Conference on Computer Vision and Pattern Recognition*, pages 1438–1447, 2019.

[37] Chuan Li and Michael Wand. Combining markov random fields and convolutional neural networks for image synthesis. In *Proceedings of the IEEE conference on computer vision and pattern recognition*, pages 2479–2486, 2016.

[38] Ian Goodfellow, Jean Pouget-Abadie, Mehdi Mirza, Bing Xu, David Warde-Farley, Sherjil Ozair, Aaron Courville, and Yoshua Bengio. Generative adversarial nets. *Advances in neural information processing systems*, 27:2672–2680, 2014.

[39] Zhaoyi Yan, Xiaoming Li, Mu Li, Wangmeng Zuo, and Shiguang Shan. Shift-net: Image inpainting via deep feature rearrangement. In *Proceedings of the European conference on computer vision (ECCV)*, pages 1–17, 2018.

[40] Guilin Liu, Fitsum A Reda, Kevin J Shih, Ting-Chun Wang, Andrew Tao, and Bryan Catanzaro. Image inpainting for irregular holes using partial convolutions. In *Proceedings of the European Conference on Computer Vision (ECCV)*, pages 85–100, 2018.

[41] Chaohao Xie, Shaohui Liu, Chao Li, Ming-Ming Cheng, Wangmeng Zuo, Xiao Liu, Shilei Wen, and Errui Ding. Image inpainting with learnable bidirectional attention maps. In *Proceedings of the IEEE/CVF International Conference on Computer Vision*, pages 8858–8867, 2019.

[42] Hongyu Liu, Bin Jiang, Yibing Song, Wei Huang, and Chao Yang. Rethinking image inpainting via a mutual encoder-decoder with feature equalizations. In *Computer Vision–ECCV 2020: 16th European Conference, Glasgow, UK, August 23–28, 2020, Proceedings, Part II 16*, pages 725–741. Springer, 2020.

[43] Şaban Öztürk. Image inpainting based compact hash code learning using modified u-net. In *2020 4th International Symposium on Multidisciplinary Studies and Innovative Technologies (ISMSIT)*, pages 1–5. IEEE, 2020.

[44] Dongsheng Wang, Chaohao Xie, Shaohui Liu, Zhenxing Niu, and Wangmeng Zuo. Image inpainting with edge-guided learnable bidirectional attention maps. *arXiv preprint arXiv:2104.12087*, 2021.

[45] Nian Cai, Zhenghang Su, Zhineng Lin, Han Wang, Zhijing Yang, and Bingo Wing-Kuen Ling. Blind inpainting using the fully convolutional neural network. *The Visual Computer*, 33(2):249–261, 2017.

[46] Yanhong Zeng, Jianlong Fu, Hongyang Chao, and Baining Guo. Learning pyramid-context encoder network for high-quality image inpainting. In *Proceedings of the IEEE conference on computer vision and pattern recognition*, pages 1486–1494, 2019.

[47] Yuqing Ma, Xianglong Liu, Shihao Bai, Lei Wang, Dailan He, and Aishan Liu. Coarse-to-fine image inpainting via region-wise convolutions and non-local correlation. In *IJCAI*, pages 3123–3129, 2019.

[48] Alec Radford, Luke Metz, and Soumith Chintala. Unsupervised representation learning with deep convolutional generative adversarial networks. *arXiv preprint arXiv:1511.06434*, 2015.

[49] Chao Yang, Xin Lu, Zhe Lin, Eli Shechtman, Oliver Wang, and Hao Li. High-resolution image inpainting using multi-scale neural patch synthesis. In *Proceedings of the IEEE conference on computer vision and pattern recognition*, pages 6721–6729, 2017.

[50] Jiahui Yu, Zhe Lin, Jimei Yang, Xiaohui Shen, Xin Lu, and Thomas S Huang. Generative image inpainting with contextual attention. In *Proceedings of the IEEE conference on computer vision and pattern recognition*, pages 5505–5514, 2018.

[51] Ugur Demir and Gozde Unal. Patch-based image inpainting with generative adversarial networks. *arXiv preprint arXiv:1803.07422*, 2018.

[52] Jiahui Yu, Zhe Lin, Jimei Yang, Xiaohui Shen, Xin Lu, and Thomas S Huang. Free-form image inpainting with gated convolution. In *Proceedings of the IEEE/CVF International Conference on Computer Vision*, pages 4471–4480, 2019.

[53] Hongyu Liu, Bin Jiang, Yi Xiao, and Chao Yang. Coherent semantic attention for image inpainting. In *Proceedings of the IEEE/CVF International Conference on Computer Vision*, pages 4170–4179, 2019.

[54] Tingzhu Sun, Weidong Fang, Wei Chen, Yanxin Yao, Fangming Bi, and Baolei Wu. High-resolution image inpainting based on multi-scale neural network. *Electronics*, 8(11):1370, 2019.

[55] Yuantao Chen, Haopeng Zhang, Linwu Liu, Xi Chen, Qian Zhang, Kai Yang, Runlong Xia, and Jingbo Xie. Research on image inpainting algorithm of improved gan based on two-discriminations networks. *Applied Intelligence*, 51(6):3460–3474, 2021.

[56] SM Uddin and Yong Ju Jung. Global and local attention-based free-form image inpainting. *Sensors*, 20(11):3204, 2020.

[57] Yuan Zeng, Yi Gong, and Jin Zhang. Feature learning and patch matching for diverse image inpainting. *Pattern Recognition*, page 108036, 2021.

[58] Yanhong Zeng, Jianlong Fu, Hongyang Chao, and Baining Guo. Aggregated contextual transformations for high-resolution image inpainting. *arXiv preprint arXiv:2104.01431*, 2021.

[59] Shengyu Zhao, Jonathan Cui, Yilun Sheng, Yue Dong, Xiao Liang, Eric I Chang, and Yan Xu. Large scale image completion via co-modulated generative adversarial networks. *arXiv preprint arXiv:2103.10428*, 2021.

[60] Min-cheol Sagong, Yong-goo Shin, Seung-wook Kim, Seung Park, and Sung-jea Ko. Pepsi: Fast image inpainting with parallel decoding network. In *Proceedings of the IEEE Conference on Computer Vision and Pattern Recognition*, pages 11360–11368, 2019.

[61] Yong-Goo Shin, Min-Cheol Sagong, Yoon-Jae Yeo, Seung-Wook Kim, and Sung-Jea Ko. Pepsi++: Fast and lightweight network for image inpainting. *IEEE Transactions on Neural Networks and Learning Systems*, 2020.

[62] Yuantao Chen, Linwu Liu, Jiajun Tao, Runlong Xia, Qian Zhang, Kai Yang, Jie Xiong, and Xi Chen. The improved image inpainting algorithm via encoder and similarity constraint. *The Visual Computer*, pages 1–15, 2020.

[63] Yi Jiang, Jiajie Xu, Baoqing Yang, Jing Xu, and Junwu Zhu. Image inpainting based on generative adversarial networks. *IEEE Access*, 8:22884–22892, 2020.

[64] Weiwei Cai and Zhanguo Wei. Piigan: Generative adversarial networks for pluralistic image inpainting. *IEEE Access*, 8:48451–48463, 2020.

[65] Chu-Tak Li, Wan-Chi Siu, Zhi-Song Liu, Li-Wen Wang, and Daniel Pak-Kong Lun. Deepgin: Deep generative inpainting network for extreme image inpainting. *arXiv preprint arXiv:2008.07173*, 2020.

[66] Johannes Fürnkranz, PK Chan, Susan Craw, Claude Sammut, William Uther, Adwait Ratnaparkhi, and L De Raedt. Mean squared error. *Encyclopedia of Machine Learning*, 2010.

[67] Deepak S Turaga, Yingwei Chen, and Jorge Caviedes. No reference psnr estimation for compressed pictures. *Signal Processing: Image Communication*, 19(2):173–184, 2004.

[68] Dragomir Anguelov, Carole Dulong, Daniel Filip, Christian Frueh, Stéphane Lafon, Richard Lyon, Abhijit Ogale, Luc Vincent, and Josh Weaver. Google street view: Capturing the world at street level. *Computer*, 43(6):32–38, 2010.

[69] Bolei Zhou, Agata Lapedriza, Aditya Khosla, Aude Oliva, and Antonio Torralba. Places: A 10 million image database for scene recognition.

IEEE transactions on pattern analysis and machine intelligence, 40(6):1452–1464, 2017.

[70] Jia Deng, Wei Dong, Richard Socher, Li-Jia Li, Kai Li, and Li Fei-Fei. Imagenet: A large-scale hierarchical image database. In *2009 IEEE conference on computer vision and pattern recognition*, pages 248–255. Ieee, 2009.

[71] David Martin, Charless Fowlkes, Doron Tal, and Jitendra Malik. A database of human segmented natural images and its application to evaluating segmentation algorithms and measuring ecological statistics. In *Proceedings Eighth IEEE International Conference on Computer Vision. ICCV 2001*, volume 2, pages 416–423. IEEE, 2001.

[72] Seyyed Hossein Soleymani and Amir Hossein Taherinia. High capacity image steganography on sparse message of scanned document image (smsdi). *Multimedia Tools and Applications*, 76(20):20847–20867, 2017.

[73] Ziwei Liu, Ping Luo, Xiaogang Wang, and Xiaoou Tang. Large-scale celebfaces attributes (celeba) dataset. *Retrieved August*, 15(2018):11, 2018.

[74] Xiuxia Cai and Bin Song. Semantic object removal with convolutional neural network feature-based inpainting approach. *Multimedia Systems*, 24(5):597–609, 2018.

13

Hybrid Leaf Generative Adversarial Networks Scheme for Classification of Tomato Leaves—Early Blight Disease Healthy

Sri Silpa Padmanabhuni[1], Pradeepini Gera[2], and A. Mallikarjuna Reddy[3]

[1,2]Department of Computer Science and Engineering, Koneru Lakshmaiah Education Foundation, Vaddeswaram, AP, India
[3]Department of Computer Science and Engineering, Anurag University, Hyderabad, Telgana, India
E-mail: silpa.padmanabhuni@gmail.com; pradeepini_cse@kluniversity.in; mallikarjunreddycse@cvsr.ac.in

Abstract

Even though India is a developing country, many countries worldwide depend on India for food because agriculture is the backbone of India. Nevertheless, farmers cannot receive productivity due to a lack of resources and awareness. So, the proposed paper focuses on smart agriculture to increase the productivity of the tomato crop by identifying the leaf diseases at an early stage by using "Hybrid GAN." The proposed deep learning classifies healthy and early blight diseases using customized GAN architecture. In general, traditional GANs are used to increase the size of the dataset. However, the proposed model uses the concept of transfer learning and customizes the last layers of neural networks to perform accurate classifications. The model also extracts its features using enhanced autoencoder neural networks. Most of the existing systems apply traditional deep learning techniques with the help of pre-trained models, which are expensive and the farmer cannot afford it. The proposed paper simplifies the modification process in the layers and reduces the deployment cost. Usage of deep learning techniques at each stage of design made the model achieve 99.9

Keywords: Basic image operations, hybrid leaf generative adversarial networks, generator, discriminator, synthetic, enhanced auto encoder, transfer learning, binary classifier.

13.1 Introduction

The proposed model in this chapter classifies the tomato leaves either as "Early Blight" or "Healthy" using the integrated GAN and transfer learning approach. Checking the plants manually to identify diseases during cultivation is complex, so the farmers need an automated system that helps in the disease detection process. The proposed model initially increases the size of the dataset to contain the images of both controlled and uncontrolled conditions; then, these images are passed to the encoder phase to extract the features from the images. Finally, the images are classified using the transfer learning approach. In designing machine learning and deep learning algorithms, the process of hyper-turning parameters plays a vital in proving the state of art model of a system. The number of parameters turning will also increase with the number of estimators. A considerable amount of data is required to balance the parameters turning efficiently. It is a tedious task to collect the data manually to simplify the process; neural networks provide different "data augmentation" [1] techniques to increase the size of the dataset, as shown in Figure 13.1.

The model can increase the size of the dataset using either basic image manipulation techniques or GANs approach in deep learning mechanisms. In this chapter, the model discusses the GANs using neural network techniques.

1. Neural Network Techniques: These techniques provide adversarial preparation, which is critical in models in which Gaussian noise does inject, and images are transformed with worst-case perturbations, resulting in incorrect answers with high confidence values. It also includes a neural style transferring technique that aids in the creation of new images by blending images, mainly when the model contains content or style reference images. It employs two distinct distance roles, one for identifying content differences and the other for identifying style differences. In GANs, the generator takes a fixed-length vector as input and outputs a domain-specific sample. Discriminator [2] takes the

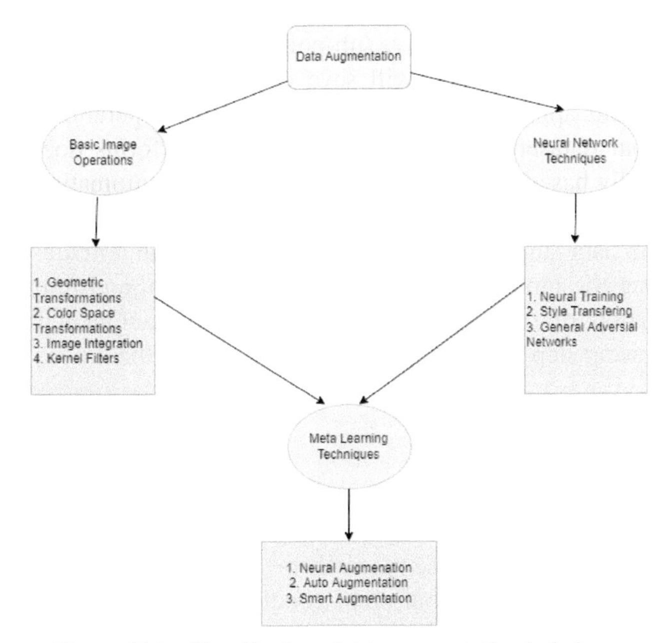

Figure 13.1 Classification of data augmentation techniques.

domain's input and generates a binary value dependent on the prediction. Generative adversarial networks, in which the task of the generator is to convert the latent space into data instances. The main goal of the generator is to create fake images so that the discriminator should be in ambiguity to decide whether the image is original or fake. The model developed a loss function to penalize the generator module if any failures occur because of the discriminator's correct prediction of a fake image. If a GAN [3] generator can build the fake images from the noisy data and tries to convert them into real images, then it is a successful GAN. The discriminator obtains the data from the original training data and fake images from the generator module [4]. The discriminator acts as the classifying neural network. It computes the loss function to identify the misclassified images; based on the loss value. It updates the weights and back propagates to the generator module to improve the module's performance.

2. Meta-learning techniques: These techniques are required to balance data between the real and virtual worlds. Google recently developed a novel method called "AutoAugment" [5], which is a policy-based approach to determining the best augmentation for a dataset [6]. The AutoAugment's

search space does divide into five sub-policies, each with two operations. This AutoAugment method will save us much time by allowing us to review and apply all possible combinations of operations used to increase our dataset's size. The operation that performs does choose automatically based on the dataset we provide; this automatic selection of operations is obtained through reinforcement learning [7]. The main goal of the data augmentation technique is to develop a model capable of handling the input that is unknown to the system and improve the generalization of the training model. The image augmentation is a pre-processing step in the deep learning area to train the model.

13.2 Literature Review

Using machine learning and deep learning techniques, several research scholars and scientists are constantly working on various plant diseases to improve production quality. These researchers used the PlantVillage dataset to obtain plant images under controlled conditions. The use of computer vision and deep learning techniques have resulted in high-accuracy systems for automatically detecting plant diseases.

Quan Huu Cap *et al.* [7] describe LeafGAN, a generative adversarial network that functions as a data augmentation tool by generating new diseased images from healthy images. This method retains the image's context while still producing high-quality images. LFLSeg, a label-free and weakly supervised segmentation module, is at the heart of this model. This model employs feature maps to extract segmentation information [8], which aids the model in implicitly learning about dense and interior regions of the leaf images. The segmented image's output does predict as a heat map, which measures the likelihood of each pixel in the final decision. Finally, this model has improved the diagnostic method and solved the issues that occur due to data overfitting.

Nazki *et al.* [9] discussed a solution to the problem of imbalanced data shifting, pipelined generative adversarial networks and did propose for plant disease detection. Two main components make this model work. The first part is AR-GAN, which does use to synthetically generate data augmentation by converting an image from one domain to another. A discriminator is a parameter that determines whether or not an image belongs to a specific domain. The image is reconstructed based on the data flows and loss functions, and its output is improved over cycle GAN [10]. For feature extraction, AR-GAN has a network with an activation reconstruction. The second part is the RESNET-50 CNN for disease detection, which has the same configuration as the baseline and uses RELU as an activation mechanism. ImageNet pre-trained weights did use to fine-tune the network.

Ho *et al.* [11] proposed a population-based augmentation algorithm with a complex strategy. The main aim of this algorithm is to use schedule learning policies to optimize the hyperparameter. After running a gradient descent algorithm on each epoch, the PBA evaluates the validation data. The weights did use to pick the bottom 25% of the data. The hyperparameters did use to select the top 25% of the data.

Lim *et al.* [12] proposed the Fast AutoAugment Algorithm based on density matching. The model creates a search space for the images and determines the calling probability and magnitude as two efficient operations. The probability distribution did apply to a pair of training datasets, the values were parameterized, and the accuracy and loss values were calculated. The efficiency is calculated using K-fold stratified shuffling and comparing the amount of identical data in both datasets. Finally, the policy does investigate using Bayesian optimization and a kernel density estimator.

Meng *et al.* [13] implemented a new way of data augmentation called DCGAN. In the olden days, data was collected manually and by rotation, flip, image brightness, etc. However, now with the help of AI and ML, the case is different. It became easy to collect the data and pass it to the CNN to identify the disease of the leaves. GAN proposed before to recognize images with the disease and improve the training and testing of the images. With the method they proposed, i.e., DCGAN, the generated images were larger but diverse. Furthermore, the images generated by this model had high quality and were easy for t-SNE and testing [14]. It would reduce the cost of time and money. They had shown that the results of this method could generate exact disease and health of the leaves. Combining the DCGAN with the GoogleNet would extract the real data that was sent as the input to the CNN [15]. And then, this CNN was trained to converge the best values for the diseased leaves.

Saranya *et al.* [16] took up a survey on the efficiency of deep learning in tomato leaf disease. They had taken up to 38 research papers for their survey. They even compared the different ways of deep learning methods that were being used in the research. They had declared that the deep learning architecture called CNN was the best algorithm used to detect the tomato leaf disease. CNN would take the image of the leaf and then send the image to the CNN architectures like AlexNet, GooglNet, VGGNet, and many more for the best results [17]. They said that the best way for data augmentation was GAN and its future algorithms. They would present the best and real-time values, so there would be no duplicates. They even said that the traditional ways of deep learning, such as KNN, ANN, SVM, GA, would not be much efficient compared to the new ways. Even in the CNN architecture, the highest accuracy one was ResNet.

Ahmed *et al.* [18] implemented a new way of an approach called MobileNetV2. As the traditional approach would take a lot of money and time, they had implemented this way for low cost for the farmers. It was a real-life application for low-cost and end devices. Farmers could not understand the technical terms; they implemented this device for the farmers. This model would not need any technical knowledge, and even farmers do not need to depend on anyone. CNN is the best algorithm; they had used deep CNN in this research for better accuracy and low-cost consumption. MobileNetV2 was the lightweight and fast architecture for the tomato leaf disease classification. Here they used contrast limited adaptive histogram equalization technique to know the disease-affected spots and lighting conditions. Usually, image to noise method was used. With this device, farmers can quickly know about the disease and take precautions.

Agarwal *et al.* [19] implemented CNN methods to detect tomato leaf disease. Here they had taken different ways to get the accurate proposed model. They used CNN method as it is the best algorithm that had been in the recent models. They took some pertained models like VGG16, InceptionV3, and MobileNet into consideration and compared them. They proposed a model that gives them better accuracy and even a low cost and time. Here, tomatoes are the most used vegetables in India, and the disease in it would cause many problems in human life like cancer and more. So the farmers need a better method to identify the disease that has affected the leaf to prevent them. The model they proposed had better accuracy and even better storage space. Here, the different augmentation techniques were applied to the proposed model to get the accuracy rate and to know the testing accuracy for all present models. They compared with all and showed that the proposed model was the best.

Gomaa *et al.* [20] proposed CNN and GAN model for TMV infection. They took this to control the disease in the early stage. The model is applied separately to each plant, taken the real-time images, and given as input to the CNN to do future work. The CNN takes the input and identifies the disease. Here they had taken these models to predict the disease in the early stage. They had taken two stages of the disease, the early stage, where the farmer can predict the disease and give the correct medicine to prevent the disease.

Table 13.1 Existing approaches with their limitations.

S.No	Author name	Algorithms used	Merits	Demerits
1	Jun Meng	DCGAN, CNN	DCGAN can be optimized according to the size and the samples.	The model can find better accuracy with the data augmentation method.
2	Mohana Saranya	CNN	The survey was taken to know which was the best architecture.	Model can apply better approaches like transfer learning and image augmentation.
3	Sabbir Ahmed	MobileNetV2, CNN	The new method is low in cost and very effective.	Model can consider varying backgrounds for better work in the algorithm.
4	Mohit Agarwal	CNN customized	The comparison is made to know a better algorithm.	Model can use pre-trained algorithms for better accuracy.
5	Ahmed Ali Gomaa	CNN, GAN	The model gets a huge amount of real-time data augmentation.	Only one disease was taken into consideration.

The second one was the diseased one where the leaf would be diseased more, and the farmers had to take precautions so that the disease was not spread. They had taken the TMV real-time dataset as their main thing and done the project over it. This model would help the farmer increase quality, local marketing, and international exporting. Even they showed the high accuracy rate obtained by combining CNN with GAN. Table 13.1 studies the limitations in the existing models to identify the gaps.

13.3 Proposed Methodology

In traditional approaches for binary classification of images, machine learning approaches did apply, but the proposed model uses neural network design. The entire process of classification is shown in Figure 13.2.

13.3.1 Data augmentation using hybrid GAN

The first phase of the proposed system uses hybrid leaf generative adversarial networks (HLGAN) to perform the data augmentation process. The GAN's composed of two models (generator and discriminator) to analyze the variations [21] in between real images and generated images. The generator module neural network learns to create the fake images by taking the random images as input. The discriminator acts as a classifier to distinguish between the original and generated images. This module computes the loss and back

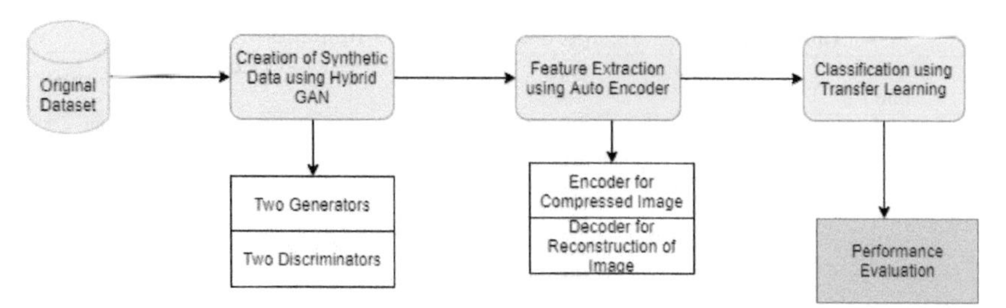

Figure 13.2 Block diagram for classification process.

propagates the values as feedback to the generator. The GAN's equation does represent in eqn (13.1). The major goal of this equation is to minimize the loss generated by the generator module and maximize the rewards of the discriminator module [22].

$$Gen_{\min}(Dis_{\max}(R(Dis, Gen)) = Gen_{\min}(Dis_{\max}$$
$$(E_{R\sim PDATA(R)} * [\log Dis(R)] + E_{N\sim P(N)} * [1 - \log Dis(Gen(N))]), \tag{13.1}$$

where $\text{Dis}_{\max}$ represents discrimination with maximization objective function.

$\text{Gen}_{\min}$ represents generator with minimization objective function.

log $(Dis(R))$ represents the discriminator loss function for real images.

1–log $Dis(Gen(N))$ represents the discriminator loss function for fake images.

In the process of augmentation, the first step is to select the layers of GAN. The proposed model uses two discriminator and two generator components of the GANs as represented in Figure 13.3.

Two-dimensional Convolutional takes an image as input and passes it through a kernel filter to calculate the dot product. The weighted matrix is converted into a feature matrix [19]. These features are important to find the nearest weights to identify the region of interest. This model helps to reduce the number of pixel operations to perform. In the proposed paper, the model uses padding with values as the same, which determines the number of pixels added to the image and adds layers with pixel values as zeros. In general, the pixels stored in the center use more than corners and edges. So to preserve the edge's information, padding helps a lot. The value same for the attribute padding illustrate as follows:

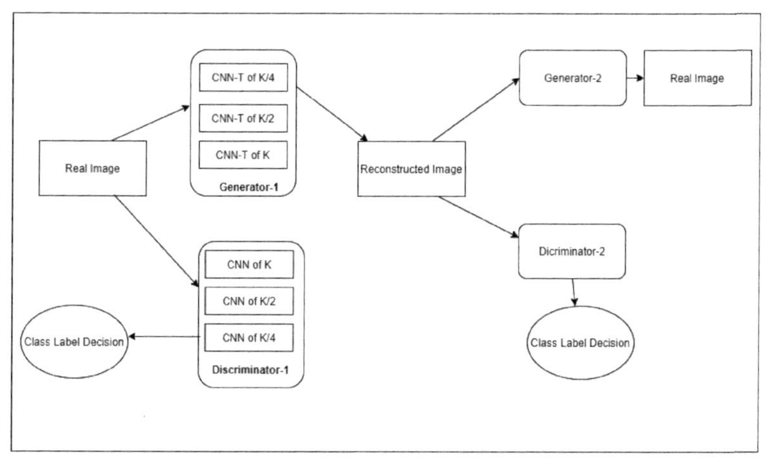

Figure 13.3 Architecture for proposed hybrid leaf GAN.

Let the number of layers be added as the border of an image $(w{\times}w)$ and the image size $(m{\times}m)$. After padding the image, the size of the image becomes as $(m + 2w)$. $(m + 2w)$. Let the kernel filter size be $(y{\times}y)$; then, the output image size is $(m + 2w - y + 1)$ after applying the kernel filter. The padding maintains the same size as of input image by using the eqn (13.2)

$$[(m + 2w)(m + 2w)] * (y \times y) \to (m \times m), \tag{13.2}$$

$where\ w = (y - 1)/2.$

The transfer function or activation function determines the output of the node in the neural network [23]. This model converts the corresponding output into binary values. The proposed paper uses Leaky ReLU, a nonlinear transfer function, which gives the output the maximum input value. It is a widely used transfer function because it implements backpropagation, and as well as it does not activate all the values simultaneously. The equation of the function represented in eqn (13.3)

$$\begin{aligned} f(x) &= 0,\ if\ x\ is\ negative\ value \\ &= x, if\ x\ is\ zero\ or\ positive\ value. \end{aligned} \tag{13.3}$$

The augmentation techniques can create different feature maps, and sometimes lower resolution pixels may contain important structural elements. So, these issues are solved by designing pooling layers. The pooling layers create a subset of feature maps for each operation separately. It detects the features irrespective of augmentation and noises contained in the image. In the proposed paper, the model used the max pooling layer. It takes the maximum value that occurs in the filter applied. The main advantage of pooling in smaller datasets is that it can avoid the overfitting problem by implementing the dropout regularization mechanism [24]. For having a good performance, the dropout value for the hidden layers can be between 0.5 and 0.8.

13.3.2 Enhanced auto encoder and decoder

Traditional feature extraction techniques work on the principle of the linear relation between the variables, but autoencoders are useful to find the nonlinear relation. The autoencoders are prominent for the automatic extraction of features where dynamic location calculations are involved. AutoEncoder is an unsupervised artificial neural network that attempts to encode the data by compressing it into the lower dimensions (bottleneck layer or code) [25] and then decoding the data to reconstruct the original input. The bottleneck layer (or code) holds the compressed representation of the input data. The model has identified a few gaps with the existing AED mechanism, which are listed as follows:

1. It is used to recreate images, but the images produced are not efficient enough when dealing with compressed images.
2. Another restriction is that the invert space vectors are not consistent. This system implies that the model can often imitate the yield pictures to include pictures.

This model is the place where enhanced autoencoders work superior to standard auto encoders. Autoencoders are similar to principal component analysis in the implementation of Feature Extraction [26]. In the proposed research, the autoencoder is designed using the estimators selected by the standard traditional values. The neural network is divided into two parts: encoder and decoder. This neural network is a self-supervised design since the input and output are the same types of images. Encoding part of autoencoders helps to learn important hidden features present in the input data to reduce the reconstruction error. During encoding, a new set of combinations of original features is generated. The design of the model is shown in Figure 13.4.

The Conv2d layers form a filter, which stores the obtained tensor values as a vector; these filters are useful for creating different texture images like sharpening, blur, detection of edges and corners, and emboss [15]. The

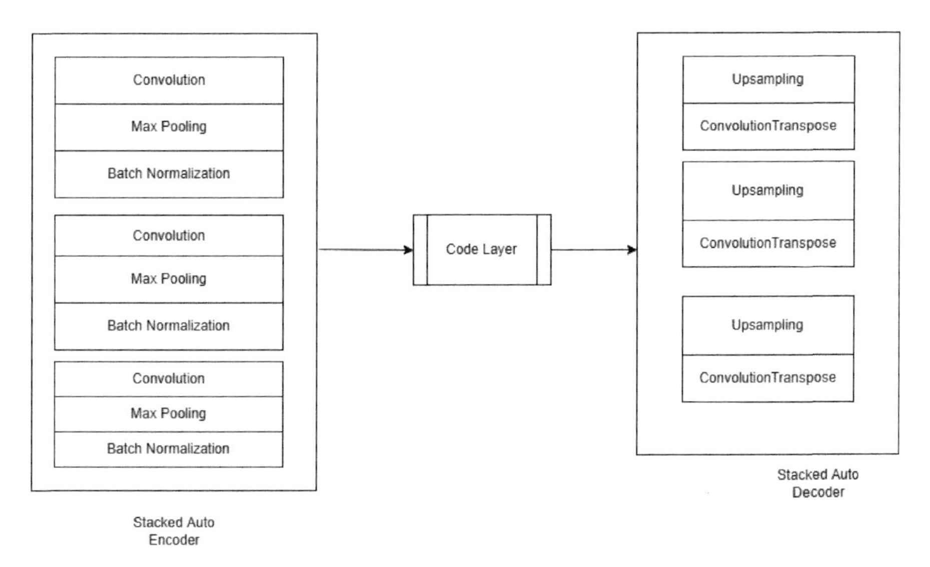

Figure 13.4 Layered architecture for auto encoder and decoder.

learning rate of this layer depends on the number of filters applied. Figure 13.5 represents the output of the conv2d layer for the proposed model.

The proposed model used a pooling layer to solve the problem of sensitivity, which occurred due to intermediate results produced by the feature map. It downsamples the features by considering only the active features present in the network. Figure 13.6 represents the output of the max pooling layer.

The output for the complete encoding and decoding process is presented in Figure 13.7, where the encoder creates the compressed image for the original data and the decoder reconstructs the images without loss of information from the compressed images.

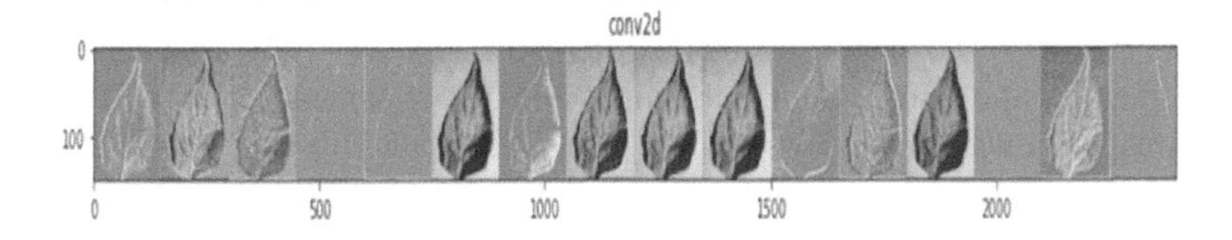

Figure 13.5 Output for 2D convolution layer.

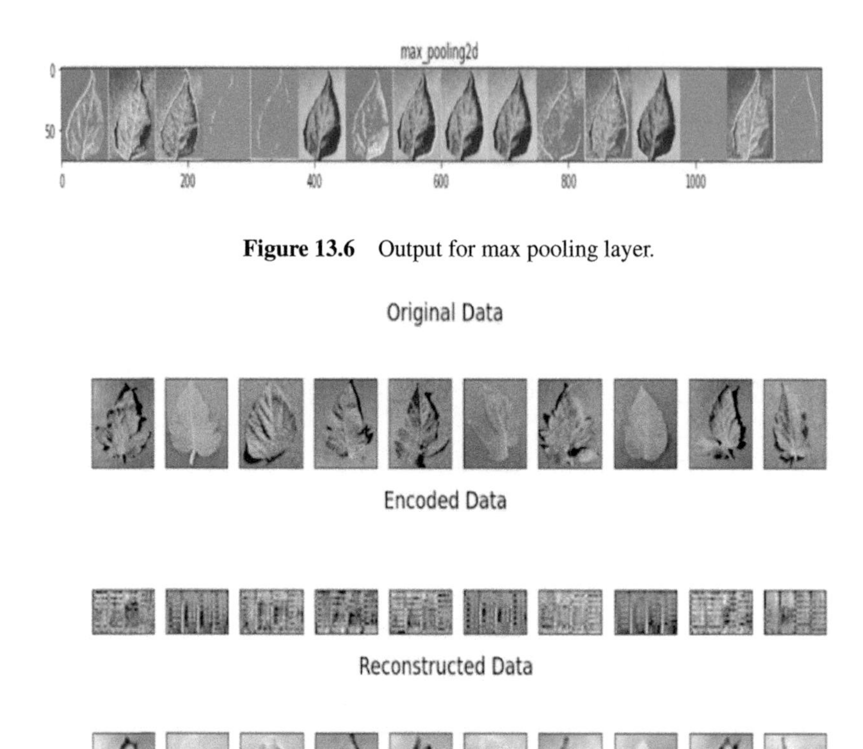

Figure 13.6 Output for max pooling layer.

Figure 13.7 Output for complete enhanced auto encoder and decoder.

13.3.3 Binary classification through transfer learning

Transfer learning initially considers a pre-trained model as a base; then, it retrains the model based on the requirement. It optimizes the estimators of the network while retraining the model. In this approach, out of the available models, the proposed paper selected "Microsoft ResNet" [27] as a base model, then it is fine-tuned to map the input and output pairs accurately. Till 2015, ImageNet was the best pre-trained model [28], but with the increase in depth of layers, it became saturated. This problem is solved by using residual blocks, which can backpropagate with updated weights when a saturation point occurs in ResNet-50—the proposed model. The hyper tuning of the model is illustrated in Table 13.2.

Algorithm 1 Classification of Early Blight and Healthy Leaves

Require: :Load the original dataset from the plant village repository

Ensure: : Classification of leaves along with their prediction values

```
1: for i in len(os_dir(tomato_leaves)) : do
2:     Train_data ← ImageDataGenerator(rescale = 1/255)
3:
4: end for                                   ▷ /*Module for Augmentation Technique*/
5: define a sequential network and initialize n = 3, m = 64, s = 3
6: for i in len(Train_data) : do
7:     for j in n : do
8:         a.Conv2D(m/2, (s, s), activation = "relu")
9:         b.Maxpooling2D()
10:        c.Dropout(0.5)
11:    end for
12:    m = m − 32
13: end for
14: for img, a in zip(img_arr, axes) : do
15:    a.a.show(img)
16:    b.a.axis("on")
17:    c.aug_images ← next(Train_data[img][a])
18: end for
19: plt.tight_layout()                        ▷ /* Module for Feature Extraction*/
20: for k in len(aug_images) : do
21:    for l in n : do
22:        a.encode_data ← Conv2D(m, (s, s), activation =' tanh', padding ='
   same')(aug_images)
23:        b.encode_data_update ← MaxPooling2D()(encode_data)
24:        c.encode_data_update ← BatchNormalization()(encode_data_update)
25:    end for
26:    m = m − 16
27: end for
28: for x in len(encode_data_update) : do
29:    for l in n : do
30:        a.decode_data ← Conv2D(m, (s, s), activation =' relu', padding ='
   same')(encoded_data_update)
31:        b.decode_data_update ← UpSampling2D((s, s))(decode_data)
32:
33:    end for
34: end for                                   ▷ /* Module for Classification*/
35: for y in len(decode_data_update) : do
36:    for z in n : do
37:        a.Classifier ← Dense(1, activation =' sigmoid'))
38:        b.Classifier_update ← compile(loss =' binary_crossentropy', optimizer =
   optimizer, metrics = ['accuracy'])
39:    end for
40: end for
41: print Classifier_update
```

13.4 Experimental Results

13.4.1 Database used

The proposed research uses the images of tomatoes collected from the open-source PlantVillage dataset [29], which is freely available in the Kaggle repository. The dataset contains images of different diseases on tomato plants and healthy plants. There are only 1591 healthy samples and 1000 early blight disease samples. All the images collected in the Kaggle repository are under a controlled environment, but the designed model should also be able to handle the uncontrolled conditions. The Figures 13.8 and 13.9 represent a few samples of both early blight and healthy images.

Table 13.2 Hyper-tuned estimators in ResNet-50.

S.No	Estimator name	Original value	Modified value
1	Batch_size	128	64
2	Learning rate	Values increases with the number of epochs	0.7 for all the epochs
3	Depth	6*n+2	27*n+m
4	Activation function	ReLU	Tanh+LeakyReLu
5	Batch normalization	Disabled	Enabled

Figure 13.8 Sample images for early blight leaf disease.

Figure 13.9 Sample images for healthy leaves.

13.4.2 Results and discussion

The experiments were conducted through an online IDE known as "Google Colab" the model implements Tensorflow and Keras library to execute the deep learning architectures. The results are conducted under standard estimators but with customized layers. The output for the discriminator model is represented as shown in Figure 13.10. The discriminator neural network uses a sequential and convolution 2D network for classifying the images.

The pooled feature map should be transferred into the column because the neural network can accept only the long vector of inputs. Then, a dense layer is applied to the neural network, which calculates the dot product of the input image and weighted data, i.e., kernel, and added to the bias value with the usage of the activation function to optimize the model. Finally, the model is compiled with the help of adam optimizer; it updates the network weights iteratively based on the training data. It is famous for its fastness. The epoch values do represent in the Figure 13.11.

Figure 13.12 represents the classification of leaves with their corresponding prediction rate. The average of the prediction rates is considered the model's accuracy. The model implemented binary classifier layers, predicting early blight and healthy images in the given dataset.

```
Model: "sequential"
_________________________________________________________________
Layer (type)                 Output Shape              Param #
=================================================================
conv2d (Conv2D)              (None, 75, 75, 16)        448
_________________________________________________________________
leaky_re_lu (LeakyReLU)      (None, 75, 75, 16)        0
_________________________________________________________________
batch_normalization (BatchNo (None, 75, 75, 16)        64
_________________________________________________________________
conv2d_1 (Conv2D)            (None, 38, 38, 32)        4640
_________________________________________________________________
leaky_re_lu_1 (LeakyReLU)    (None, 38, 38, 32)        0
_________________________________________________________________
batch_normalization_1 (Batch (None, 38, 38, 32)        128
_________________________________________________________________
flatten (Flatten)            (None, 46208)             0
_________________________________________________________________
dense (Dense)                (None, 1)                 46209
=================================================================
Total params: 51,489
Trainable params: 51,393
Non-trainable params: 96
_________________________________________________________________
```

Figure 13.10 Output for generator model.

```
Epoch 1/15
49/49 [==============================] - 36s 733ms/step - loss: 0.2839 - accuracy: 0.9122 - val_loss: 0.7233 - val_accuracy: 0.4000
Epoch 2/15
49/49 [==============================] - 26s 527ms/step - loss: 0.1507 - accuracy: 0.9183 - val_loss: 0.9869 - val_accuracy: 0.3900
Epoch 3/15
49/49 [==============================] - 25s 502ms/step - loss: 0.1114 - accuracy: 0.9524 - val_loss: 0.3323 - val_accuracy: 0.8400
Epoch 4/15
49/49 [==============================] - 25s 502ms/step - loss: 0.0494 - accuracy: 0.9835 - val_loss: 0.3017 - val_accuracy: 0.8500
Epoch 5/15
49/49 [==============================] - 25s 502ms/step - loss: 0.0485 - accuracy: 0.9886 - val_loss: 0.2070 - val_accuracy: 0.9200
Epoch 6/15
49/49 [==============================] - 24s 499ms/step - loss: 0.1324 - accuracy: 0.9462 - val_loss: 0.7213 - val_accuracy: 0.5300
Epoch 7/15
49/49 [==============================] - 24s 498ms/step - loss: 0.0929 - accuracy: 0.9659 - val_loss: 0.3181 - val_accuracy: 0.7900
Epoch 8/15
49/49 [==============================] - 25s 500ms/step - loss: 0.0611 - accuracy: 0.9741 - val_loss: 0.4429 - val_accuracy: 0.7900
Epoch 9/15
49/49 [==============================] - 25s 502ms/step - loss: 0.0557 - accuracy: 0.9659 - val_loss: 0.6358 - val_accuracy: 0.5800
Epoch 10/15
49/49 [==============================] - 25s 502ms/step - loss: 0.0165 - accuracy: 0.9928 - val_loss: 0.1411 - val_accuracy: 0.9200
Epoch 11/15
49/49 [==============================] - 25s 502ms/step - loss: 0.0040 - accuracy: 0.9990 - val_loss: 0.0612 - val_accuracy: 0.9700
Epoch 12/15
49/49 [==============================] - 25s 503ms/step - loss: 0.0061 - accuracy: 0.9990 - val_loss: 0.2557 - val_accuracy: 0.9100
Epoch 13/15
49/49 [==============================] - 25s 503ms/step - loss: 7.3316e-04 - accuracy: 1.0000 - val_loss: 0.3898 - val_accuracy: 0.9000
Epoch 14/15
49/49 [==============================] - 25s 507ms/step - loss: 2.6175e-04 - accuracy: 1.0000 - val_loss: 0.2725 - val_accuracy: 0.9100
```

Figure 13.11 A sample screen for execution of epochs.

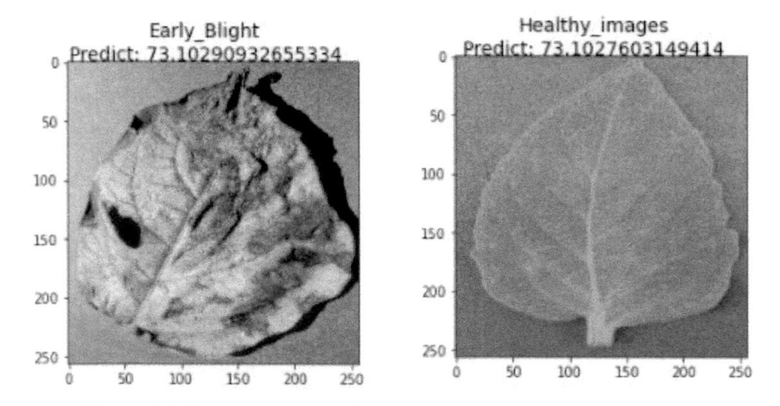

Figure 13.12 Binary classification of images using transfer learning.

The performance of the deep learning and machine learning modules are represented using various parameters like accuracy, recall, precision, and F1-score. In the proposed system, the visualization graph does represent using accuracy and loss parameters for both training and test data, as shown in the Figure 13.13.

The proposed to prove its efficiency in terms of accuracy, it has compared the few existing algorithms and presented in Table 13.3 and Figure 13.14.

In Figure 13.14, X-axis represents the name of the algorithms considered from the existing models along with the proposed algorithm, and Y-axis represents the accuracy obtained by the different models. The figure shows

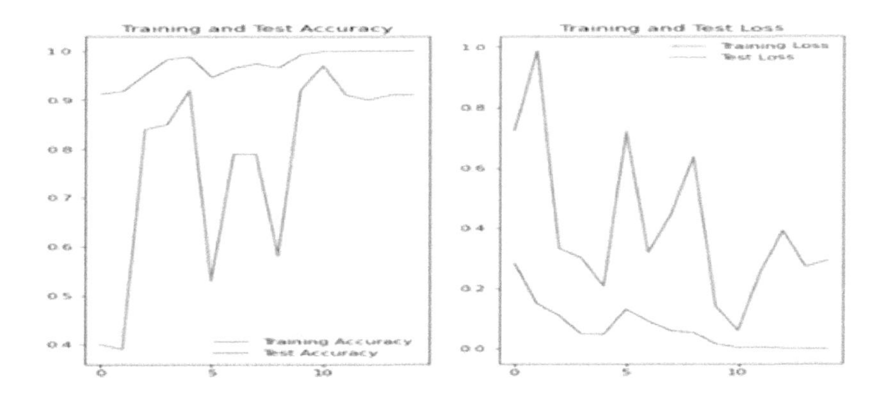

Figure 13.13 Screenshot of training, test accuracy.

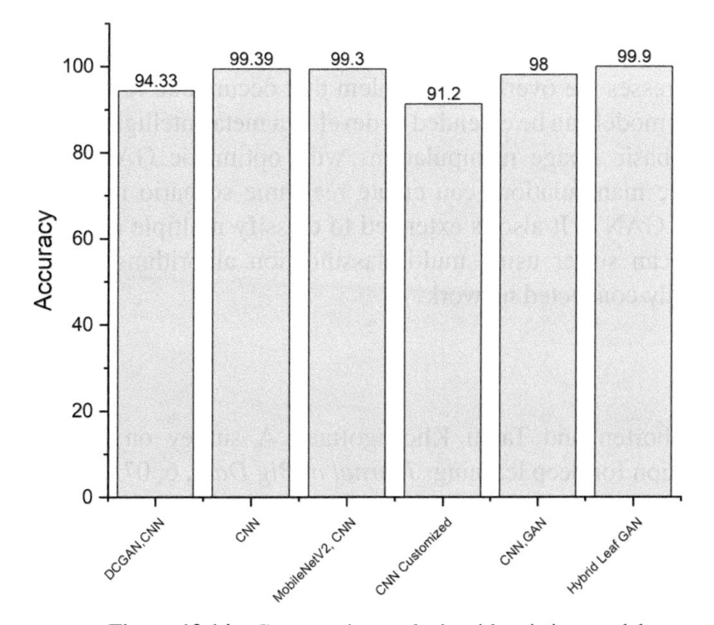

Figure 13.14 Comparative analysis with existing model.

that the proposed algorithm has obtained the highest accuracy compared with the existing ones.

Table 13.3 Efficiency of the proposed research.

S.No	Author name	Algorithms used	Accuracy
1	Jun Meng	DCGAN, CNN	94.33
2	Mohana Saranya	CNN	99.39
3	Sabbir Ahmed	MobileNetV2, CNN	99.30
4	Mohit Agarwal	CNN customized	91.2
5	Ahmed Ali Gomaa	CNN, GAN	98.0
6	Proposed	Hybrid Leaf GAN	99.9

13.5 Conclusion

In a real-time environment, the conditions are uncontrolled. The system may face problems like labeling of disease dataset, addressing the data imbalance problem, and overfitting problems because of the size of diagnostic clues, which are smaller in size than in the real-time environment. The system also suffers from the overfitting problem because it takes many computations to select and apply possible geometric transformations. So, by designing a good policy schema, like Hybrid Leaf GAN, the auto augmentation process can efficiently improve the accuracy by using the pre-trained models. The Hybrid Leaf GAN modules generate more training data because of which the system addresses the overfitting problem that occurs due to basic image operations. The model can be extended to develop a meta-intelligent approach by combining basic image manipulations with optimistic GAN in future work. The basic manipulations can create real-time scenario images more efficiently than GAN's. It also is extended to classify multiple diseases that tomato leaves can suffer using multi-classification algorithms as the last layers of the fully connected network.

References

[1] Connor Shorten and Taghi Khoshgoftaar. A survey on image data augmentation for deep learning. *Journal of Big Data*, 6, 07 2019.

[2] Huzaifa Maniyar and Suneeta Budihal. *Plant Disease Detection: An Augmented Approach Using CNN and Generative Adversarial Network (GAN)*, pages 252–261. 11 2020.

[3] Abdul Waheed, Muskan Goyal, Deepak Gupta, Ashish Khanna, Aboul Ella Hassanien, and Hari Pandey. An optimized dense convolutional neural network model for disease recognition and classification in corn leaf. *Computers and Electronics in Agriculture*, 175:105456, 08 2020.

[4] Jia-Rong Xiao, Pei-Che Chung, Hung-Yi Wu, Quoc-Hung Phan, Jer-Liang Yeh, and Max Hou. Detection of strawberry diseases using a convolutional neural network. *Plants*, 10:31, 12 2020.

[5] Rakesh Joshi, Manoj Kaushik, Malay Dutta, Ashish Srivastava, and Nandlal Choudhary. Virleafnet: Automatic analysis and viral disease diagnosis using deep-learning in vigna mungo plant. *Ecological Informatics*, 61, 11 2020.

[6] Mehmet Metin Ozguven and Kemal Adem. Automatic detection and classification of leaf spot disease in sugar beet using deep learning algorithms. *Physica A*, 535(122537):122537, December 2019.

[7] Takumi Saikawa, Quan Huu Cap, Satoshi Kagiwada, Hiroyuki Uga, and Hitoshi Iyatomi. AOP: An anti-overfitting pretreatment for practical image-based plant diagnosis. In *2019 IEEE International Conference on Big Data (Big Data)*. IEEE, December 2019.

[8] Sri Silpa Padmanabhuni. An extensive study on classification based plant disease detection system. *JOURNAL OF MECHANICS OF CONTINUA AND MATHEMATICAL SCIENCES*, 15, 05 2020.

[9] Haseeb Nazki. Synthetic data augmentation for plant disease image generation using gan. 05 2018.

[10] Hari Mohan Pandey. Intelligent classification and analysis of essential genes using quantitative methods. *ACM Transactions on Multimedia Computing, Communications, and Applications (TOMM)*, 16:1 – 21, 2020.

[11] Daniel Ho, Eric Liang, Ion Stoica, Pieter Abbeel, and Xi Chen. Population based augmentation: Efficient learning of augmentation policy schedules, 2019.

[12] Sungbin Lim, Ildoo Kim, Taesup Kim, Chiheon Kim, and Sungwoong Kim. Fast autoaugment, 2019.

[13] Qiufeng Wu, Yiping Chen, and Jun Meng. DCGAN-based data augmentation for tomato leaf disease identification. *IEEE Access*, 8:98716–98728, 2020.

[14] Padmaja Grandhe. A novel method for content based 3d medical image retrieval system using dual tree m-band wavelets transform and multiclass support vector machine. *Journal of Advanced Research in Dynamical and Control Systems*, 12:279–286, 03 2020.

[15] Mohamed Loey, Ahmed Elsawy, and Mohamed Afify. Deep learning in plant diseases detection for agricultural crops: A survey. 11:18, 02 2020.

[16] S Saranya, R Rajalaxmi, R Prabavathi, T Suganya, S Mohanapriya, and T Tamilselvi. Deep learning techniques in tomato plant- a review. *Journal of Physics: Conference Series*, 1767:012010, 02 2021.

[17] Saiyed Umer, Partha Mohanta, Ranjeet Rout, and Hari Pandey. Machine learning method for cosmetic product recognition: a visual searching approach. *Multimedia Tools and Applications*, 80, 11 2021.

[18] Sabbir Ahmed, Md. Bakhtiar Hasan, Tasnim Ahmed, Redwan Karim Sony, and Md. Hasanul Kabir. Less is more: Lighter and faster deep neural architecture for tomato leaf disease classification, 2021.

[19] Mohit Agarwal, Abhishek Singh, Siddhartha Arjaria, Amit Sinha, and Suneet Gupta. ToLeD: Tomato leaf disease detection using convolution neural network. *Procedia Comput. Sci.*, 167:293–301, 2020.

[20] Ahmed Ali Gomaa and Yasser M Abd El-Latif. Early prediction of plant diseases using CNN and GANs. *Int. J. Adv. Comput. Sci. Appl.*, 12(5), 2021.

[21] Muhammad Saleem, J. Potgieter, and Khalid Arif. Plant disease detection and classification by deep learning. *Plants*, 8:468, 10 2019.

[22] Surampalli Ashok, Gemini Kishore, Velpula Rajesh, S Suchitra, S G Gino Sophia, and B Pavithra. Tomato leaf disease detection using deep learning techniques. In *2020 5th International Conference on Communication and Electronics Systems (ICCES)*. IEEE, June 2020.

[23] S Hernández and Juan L López. Uncertainty quantification for plant disease detection using bayesian deep learning. *Appl. Soft Comput.*, 96(106597):106597, November 2020.

[24] Junde Chen, Jinxiu Chen, Defu Zhang, Yuandong Sun, and Y A Nanehkaran. Using deep transfer learning for image-based plant disease identification. *Comput. Electron. Agric.*, 173(105393):105393, June 2020.

[25] Drasko Radovanovic and Slobodan Dukanovic. Image-based plant disease detection: A comparison of deep learning and classical machine learning algorithms. In *2020 24th International Conference on Information Technology (IT)*. IEEE, February 2020.

[26] Dor Oppenheim, Guy Shani, Orly Erlich, and Leah Tsror. Using deep learning for image-based potato tuber disease detection. *Phytopathology*, 109, 12 2018.

[27] Davinder Singh, Naman Jain, Pranjali Jain, Pratik Kayal, Sudhakar Kumawat, and Nipun Batra. PlantDoc. In *Proceedings of the 7th ACM IKDD CoDS and 25th COMAD*, New York, NY, USA, January 2020. ACM.

[28] Abirami Devaraj, Karunya Rathan, Sarvepalli Jaahnavi, and K Indira. Identification of plant disease using image processing technique. In *2019 International Conference on Communication and Signal Processing (ICCSP)*. IEEE, April 2019.

[29] David P. Hughes and Marcel Salath'e . An open access repository of images on plant health to enable the development of mobile disease diagnostics through machine learning and crowdsourcing. *CoRR*, abs/1511.08060, 2015.

Index

About the Editors

Roohie Naaz Mir is a professor in the Department of Computer Science & Engineering at NIT Srinagar, INDIA. She received her B.Eng. (Hons) in Electrical Engineering from University of Kashmir (India) in 1985, her M.Eng. in Computer Science & Engineering from IISc Bangalore (India) in 1990 and Ph.D. from University of Kashmir (India) in 2005. She is a fellow of IEI and IETE India, a senior member of IEEE and a member of IACSIT and IAENG. She is the author of many scientific publications in international journals and conferences. Her current research interests include reconfigurable computing and architecture, mobile and pervasive computing, security and routing in wireless ad hoc, and sensor networks.

Vipul Kumar Sharma is working as an Assistant Professor (Grade-II) in the department of Computer Science Engineering & Information Technology, Jaypee University of Information Technology, Solan, India. He received his Ph.D. in Computer Vision with Deep Learning from National Institute of Technology Srinagar, INDIA in the year 2021. He received his B.Tech (Hons) degree in Computer Science & Engineering from Lovely Professional University, Punjab in 2011. His research interests include pattern recognition, deep learning, steganography, digital image processing, pattern recognition, and machine learning.

Ranjeet Kumar Rout is currently serving as Assistant Professor in the Department of Computer Science and Engineering, National Institute of Technology Srinagar, Hazratbal, India. He received his Ph.D. degree from the Department of Information Technology of Indian Institute of Engineering Science and Technology Shibpur, West Bengal, India in 2018. Prior to working at NIT Srinagar, Dr. Ranjeet had some useful research and teaching experience at the National Institute of Technology Jalandhar, Punjab. Dr. Ranjeet has also worked as research personnel at Indian Statistical Institute, Kolkata. His research interests include machine learning, deep learning, visual cryptography, and computational biology. He holds three patents and has published several papers in peer-reviewed international and scientific journals in the field of non-linear Boolean functions and computational biology and affective computing.

Saiyed Umer received his B.Sc. (Hons) degree in Mathematics from Vidyasagar University, India in 2005. He earned a Master of Computer Application from the West Bengal University of Technology, India in 2008, an M.Tech. from the University of Kalyani, India in 2012, and a Ph.D. from the Department of Information Technology at Jadavpur University, Kolkata, India. He was part of the research personnel at Indian Statistical Institute (ISI), Kolkata, India, from November 2012 to April 2017. Currently, he has joined as an Assistant Professor in the Department of Computer Science and Engineering, Aliah University (Govt. of West Bengal, India), Kolkata, India. His research interests include computer vision, machine learning, deep learning, and business data analytics.

9788770227025